GUIDELINES FOR
USE OF GLASS IN BUILDINGS

GUIDELINES FOR
USE OF GLASS IN BUILDINGS

Edited by
N K GARG
Professor
Department of Architecture
Faculty of Architecture, Maipal Institute of Technology
Manipal, Karnataka

NEW AGE
REFERENCE

An Imprint of

NEW AGE INTERNATIONAL (P) LIMITED, PUBLISHERS
LONDON • NEW DELHI • NAIROBI
Bangalore • Chennai • Cochin • Guwahati • Hyderabad • Kolkata • Lucknow • Mumbai
Visit us at **www.newagepublishers.com**

Published by New Age International (P) Ltd., Publishers
First Edition: 2007
Reprint: 2018

GLOBAL OFFICES

- **New Delhi** NEW AGE INTERNATIONAL (P) LIMITED, PUBLISHERS
  7/30 A, Daryaganj, New Delhi-110002, (INDIA)
  Tel.: (011) 23253771, 23253472, **Telefax:** 23267437, 43551305
  E-mail: contactus@newagepublishers.com • Visit us at www.newagepublishers.com
- **London** NEW AGE INTERNATIONAL (UK) LTD.
  27 Old Gloucester Street, London, WC1N 3AX, UK
  E-mail: info@newacademicscience.co.uk • Visit us at www.newacademicscience.co.uk
- **Nairobi** NEW AGE GOLDEN (EAST AFRICA) LTD.
  Ground Floor, Westlands Arcade, Chiromo Road (Next to Naivas Supermarket)
  Westlands, Nairobi, KENYA, **Tel.:** 00-254-713848772, 00-254-725700286
  E-mail: kenya@newagepublishers.com

BRANCHES

- **Bangalore** 37/10, 8th Cross (Near Hanuman Temple), Azad Nagar, Chamarajpet, Bangalore- 560 018
  Tel.: (080) 26756823, **Telefax:** 26756820, **E-mail: bangalore@newagepublishers.com**
- **Chennai** 26, Damodaran Street, T. Nagar, Chennai-600 017, **Tel.:** (044) 24353401, **Telefax:** 24351463
  E-mail: chennai@newagepublishers.com
- **Cochin** CC-39/1016, Carrier Station Road, Ernakulam South, Cochin-682 016, **Tel.:** (0484) 2377303, **Telefax:** 40513
  E-mail: cochin@newagepublishers.com
- **Guwahati** Hemsen Complex, Mohd. Shah Road, Paltan Bazar, Near Starline Hotel, Guwahati-781 008,
  Tel.: (0361) 2513881, **Telefax:** 2543669, **E-mail: guwahati@newagepublishers.com**
- **Hyderabad** 105, 1st Floor, Madhiray Kaveri Tower, 3-2-19, Azam Jahi Road, Near Kumar Theater, Nimboliadda Kachiguda, Hyderabad-500 027, **Tel.:** (040) 24652456, **Telefax:** 24652457
  E-mail: hyderabad@newagepublishers.com
- **Kolkata** RDB Chambers (Formerly Lotus Cinema) 106A, 1st Floor, S N Banerjee Road, Kolkata-700 014
  Tel.: (033) 22273773, **Telefax:** 22275247, **E-mail: kolkata@newagepublishers.com**
- **Lucknow** 16-A, Jopling Road, Lucknow-226 001, **Tel.:** (0522) 2209578, 4045297, **Telefax:** 2204098
  E-mail: lucknow@newagepublishers.com
- **Mumbai** 142C, Victor House, Ground Floor, N.M. Joshi Marg, Lower Parel, Mumbai-400 013
  Tel.: (022) 24927869, **Telefax:** 24915415, **E-mail: mumbai@newagepublishers.com**
- **New Delhi** 22, Golden House, Daryaganj, New Delhi-110 002, **Tel.:** (011) 23262368, 23262370, **Telefax:** 43551305
  E-mail: sales@newagepublishers.com

ISBN: 978-81-224-2065-4
C-17-10-10806

Printed in India at Glorious Printers, Delhi.
Typeset at In-house, Delhi.

NEW AGE INTERNATIONAL (P) LIMITED, PUBLISHERS
7/30 A, Daryaganj, New Delhi-110002
Visit us at **www.newagepublishers.com**
(CIN: U74899DL1966PTC004618)

Dedicated to

Central Building Research Institute, Roorkee

my mother organization which nurtured me for 35 years and provided me opportunities to work over a wide spectrum of projects including

Use of Glass in Buildings

— Dr. N. K. Garg

Foreword

Glass has always been associated with notable advances in construction. The monumental edifice of Crystal Palace, built in London by Paxton in 1851, was erected in record time using revolutionary assembly techniques. With its fully glazed roof, the "glass ceiling" concept received its seal of approval by architects. This initial and tentative use of glass on such a spectacular and large scale was contemporary to the extraordinary contribution engineering, supplemented by the industrial revolution. It was to engender the unprecedented development of construction allying glass with steel, cast iron and, in due course, concrete.

Beyond its architectural aspect & structural capabilities, glass has come up with new properties to meet the modern demands of comfort as well as safety & security. It is a rare construction material, which is traditional, multifunctional & constantly evolving.

The 20th century has witnessed increasingly complex performance demands on glass. It has lost its image for fragility and earned the status of a multi-purpose building material. More than ever, glass has become a symbol of modernity.

The glass industry in India has kept pace with the emerging performance requirements of glass. It is capable of producing world-class glass in the country. Newer performance characteristic needed guidelines for use of glass most scientifically and judiciously in buildings. I am happy to note that Central Building Research Institute has undertaken the enabling role and prepared guidelines for use of glass in buildings.

The guidelines covering general as well as optical, acoustical & structural properties of commonly used glass types will enable the designer to select the right type of glass for his building. Tools to arrive at the appropriate thickness of glass, innovative glazing systems along with do's & don'ts are also covered. The document with its holistic approach will surely promote right and safe use of glass in buildings as a symbol of modernity through the 21st century.

V.K.Mathur

Former Director, CBRI

Preface

Glass is a "high-tech" material appreciated for its visual appeal. The fascination which glass inspires in designers is explained by its aesthetics and functions. Its transparency, affording both light and vision yet imposing separation and even isolation worth appreciating. It can, do the very opposite as well. Glass can form building's envelope as opaque as stone, reflecting images and allowing light to reach the interior, without compromising vision through to the exterior. Glass can also provide a delicate covering that softens a building's mass, or indeed accentuate it, by simply reflecting both light and the surrounding environment.

The innovative installation techniques e.g. structural glazing or bolted glass assemblies and developments in glass coating technology for solar control performance has extended its capacity to constitute entire facades, forming a skin which can be totally smooth, transparent or opaque, reflective or coloured as per the perception of the designer.

India is producing world class glass with a range of strength & optical properties. To facilitate its use as an emerging building material adding excellence in building performance, guidelines have been formulated. National & international codes, experiences of the glass industry including the manufactures, the fabricators & the installers form the basis of these guidelines. It is my endeavor to provide technical information accurately. However variations are possible and CBRI shall not be liable for the same. The role of All India Flat Glass Manufacturers' Association (AIFGMA) for sponsoring this study at CBRI & providing necessary assistance is commendable.

It has been a pleasant experience for me to edit this volume. Sh. V. K. Mathur, Former Director, CBRI has very kindly written the foreword. I am indebted to him. I appreciate the dedication of my colleagues: Sh. Ashok Kumar, Sh. S.K.Negi, Sh. Navjeev Saxena, Dr. Rajesh Deoliya and Sh. V. Srinivasan at CBRI in completing the project entitled, 'Use of Glass in Buildings' which forms the basis of this work.

(Dr. N. K. Garg)

Acknowledgements

Eminent industrialists, manufacturers, fabricators, installers, administrators, scholars & users of glass have provided their inputs related to their fields of expertise. Special recognition is due to the following for facilitating the preparation of guidelines, 'use of glass in buildings'.

Shri K. K. Trivedi
Secretary General
All India Flat Glass Manufacturers Association
(AIFGMA)
New Delhi

Mr. Alok Modi
Jt. M. Director
Gujarat Guardian Limited
New Delhi

Shri P. K. Kheruka
Director
Borosil Glass Works Limited
Mumbai

Shri Sharan Jeet Singh
Managing Director
GSC Toughened Glass Pvt. Ltd.
Noida

Sh. Deepak Ghalowt
Architect
Editor
Construction Journal of India
New Delhi

Shri S. K. Mehrotra
V. P. (Works)
Gujarat Borosil Limited
Bharuch
(Gujarat)

Shri Phiroze J. Masters
General Manager (Mktg.)
Gujarat Borosil Limited
Mahalakshmi, Mumbai

Sh. R. K. Bhatt
Chief Engineer
DLF Constructions
Gurgoan

Shri Gopal C. Panigrahi
Head Manufacturing
Float Glass India Limited
Maharashtra

Shri R. M. Kunzru
Executive Director (Technical & Operations)
Float Glass India Limited
Taloja, Maharashtra

Mr. Farhaat Kamil
Kaenat Glass Agencies
New Delhi

Shri A. L. Kelkar
Joint General Manager
ICICI Bank Limited
Mumbai

Shri R. Madhukar
Chief Manager
ICICI Bank Limited
Mumbai

Shri Oswald Lobo
Proprietor
Professional Glass Façade Cleaners
Kandivali (E)
Mumbai

Sh. Sandeep Chaudri
Partner, A L Karma
New Delhi

Sh. Anshu Gupta
Territory Head-Arch Division
Gujarat Guardian Limited
New Delhi

Sh. Subash Chandra
Raw Line Manager
Gujarat Guardian Limited
Ankleshwar
Gujrat

Shri Anand R. Jain
Saint – Gobain Glass India Ltd.
Sriperumbudur – 602 105
Kanchipuram Distt.
Tamil Nadu

Contents

1.	**Introduction**	**1**
2.	**Types of Glass, Properties and Uses**	**7**
	2.1 Normal (Annealed) Glass	8
	2.2 Laminated Glass	11
	2.3 Tempered or Toughened Glass	13
	2.4. Heat Strengthened Glass	15
	2.5. Heat Soaked Tempered Glass	16
	2.6 Reflective Glass	17
	2.7 Insulating Glass Unit (Double Glazing)	19
	2.8 Mirror	21
3.	**Selecion of Specific Glass**	**25**
	3.1 General	26
	3.2 Heat Gain	26
	3.3 Sound Insulation	36
	3.4 Safety & Security	40
	3.5 Strength	49
	3.6 Aesthetics	49
	3.7 General Information	50
4.	**Determination of Appropriate Thickness of Glass**	**57**
	4.1 General	58
	4.2 Standard Nominal Thickness (SNT)	59
	4.3 Design Considerations	60
	4.4 Determination of Minimum Glass Thickness	61
	4.5 Net Design Wind Pressure (P_{net})	64
	4.6 Examples to Determine the Appropriate Thickness of Glass Panel for an Office Building in Delhi Region.	70
5.	**Glazing Systems**	**103**
	5.1 Curtain Walling	104
	5.2 Structural Glazing	104

5.3 Bolted Glazing 105
5.4 Fin Supported Glazing 106
5.5 Cable Stayed Glazing 106
5.6 Suspended Glazing 107
5.7 Mirror Installation 108
5.8 Accessories for Glazing Systems 109

6. General Guidelines 117

6.1 General 118
6.2 Normal (Annealed) Glass 118
6.3 Laminated Glass 119
6.4 Tempered Glass 119
6.5 Reflective Glass 119
6.6 Insulating Glass 121
6.7 Mirror 121
6.8 Aluminium Framings 123
6.9 Safety from Earthquakes 124

Reference Material 125
Definitions 131

1

Introduction

1.0 INTRODUCTION

Glass is one of the commonest man-made materials that has been in use for a number of years. From the stained glass windows of medieval churches, glass has brought us protection from our environment, while also reflecting its own beauty. Today's flat glass still exhibits those ancient characteristics of form and function, while carrying us into the future as a performance platform for development of technologies. Using a wide variety of batch combinations, we take glass from our float lines and coat, bend, shape, laminate and temper it. The resulting products provide us with comfort, protect our fabrics from fading, reduce our energy costs, block sound transmission, improve our security and allow us to replace walls of brick and mortar with panoramas of light and natural beauty.

The manufacturing of glass began around 1500 B.C. in Egypt & Mesopotamia. Syrian glassmakers invented the blowpipe that enabled the production of glass in countless shapes. The Romans further developed glass making in first century A.D. to produce flat glass by blowing huge glass bubbles or cylinders, which were opened and flattened. The Venetians began their glass industry in 13th century. The manufacturing process for plate glass, which involved casting glass on iron table and then polishing it was developed. The expertise spread throughout Europe. French glassmakers improved the process by using large tables to make more sizeable pieces of glass. The molten glass was poured onto special tables & rolled out flat. After cooling, it was ground by means of rotating cast iron discs & increasingly fine abrasive sands and then polished using felt disks. The result was flat glass with good optical transmission quality. Royal Glass Works' was set up in France in 1665 to manufacture mirrors for 'Palace of Versailles'.

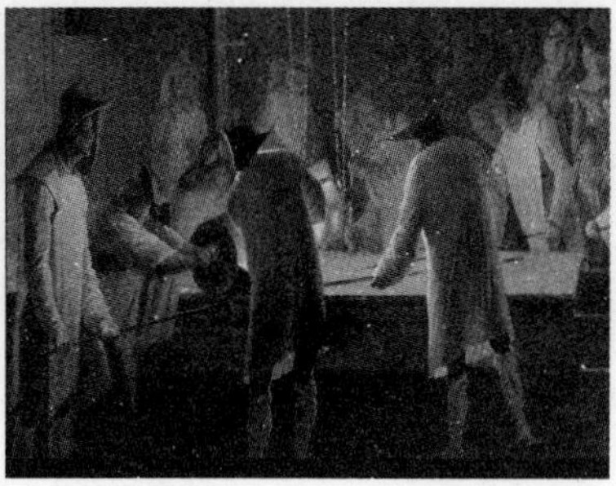

England became the world's center for quality plate glass for windows in 1773. This period marked for the first time in the history of glassmaking that glazed window glass was widely available and affordable for most homeowners. Industrial revolution brought a number of innovations to glassmaking. William Pilkington invented a machine, which automated the production of glass. Around the turn of the century, glassmakers discovered that glass could be "tempered" by reheating it and then cooling it again quickly.

The American Irwing Colburn and the Belgian Emile Fourcault simultaneously developed a new automated glassmaking process, which drew molten glass from the furnace in a thin stream, then flattened and cooled it by pulling it between asbestos rollers as a sheet. Though the sheet glass produced by this "draw" process was still not distortion free, it was the highest quality glass ever produced. A ribbon of glass is drawn vertically from the tank or furnace up a temperature controlled annealing tower. The newly formed glass emerges at the top of the tower where automatic cutters, and hydraulically controlled suckers, enable the operator to snap off the sheet glass, which is then sent to the warehouse.

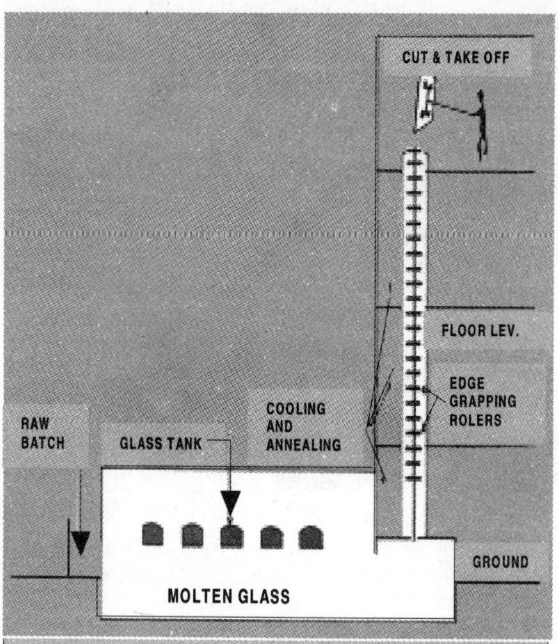

THE MANUFACTURE OF SHEET GLASS

Glass manufacturing changed forever when Alastair Pilkington invented the modern float glass in the 1950s. In this process, molten glass was poured in a continuous stream into a shallow pool of molten metal, typically tin. The molten glass would spread onto the surface of the metal producing a high quality, consistently level sheet of glass known as float glass. Pilkington process revolutionized worldwide the flat glass industry in a number of ways. It drove down the cost of glass dramatically and created new applications such as the exterior of high rise office buildings.

Float glass is manufactured by melting sand. Other ingredients such as soda ash, limestone and salt cake are added to lower the melting temperature of silica and promote optical clarity of the finished glass product. Mixed batch of above materials is heated to a temperature of about 1550°C and formed into large flat sheets by floating molten glass on molten tin thus giving it precise flatness and transparency.

The float process enabled a number of new technologies and product developments. For the first time, uniform, high quality float glass could be made in a variety of thick nesses, to address safety, security or noise reduction concerns while still meeting the highest aesthetic standards. The float process made it easy to vary the composition of the molten glass to make products for special applications such as tinted glass.

With the worldwide energy crisis in the early 1970s, the glass industry responded with new developments that answered consumers growing concern about energy

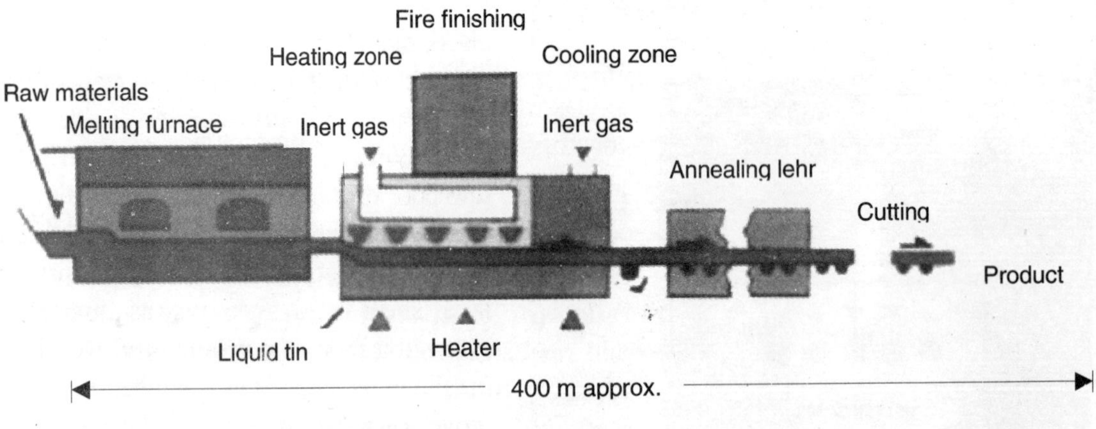

Manufacturing Process of Float Glass

efficiency and addressed new performance concerns. For example, with expanded awareness of solar energy, glassmakers developed new coatings that would help flat glass to retain passive solar and radiant heat more efficiently as well as solar control coatings that would help to block the sun's heat in warmer climates while still allowing visible light transmittance. In addition, glass manufacturers began to introduce new high and medium performance reflective coatings that would enable builders and architects to achieve specific performance characteristics in terms of visible light transmittance, solar reflectance, and shading coefficient making possible the wide range of attractive energy-efficient buildings we see around us today.

As the worldwide economy improved in the early 1980s, concerns about energy efficiency remained and glass industry continued its efforts to develop innovative new technologies and processes that would maximize window performance. Low emissivity coating is one of the biggest developments in the flat glass industry. These coatings enable architects to take advantage of passive solar heat during the winter months, while reflecting radiant heat back into the buildings interior. These products are still gaining ground in the market place, as builders and consumers learn about the advantages of "Low-E" coatings.

While the process of revolution/evolution of newer types of glasses continues, different types of glasses with different properties need to be specified for their use in buildings judiciously.

All India Flat Glass Manufacturers Association has taken a forward looking role and supported the preparation of guidelines for use of glass in commercial and residential buildings. The commonly used glass types – Normal (Annealed), Laminated, Tempered, Reflective, Insulating and Mirror are covered in this publication.

The guidelines consist of Six Chapters

Chapter 2 describes general properties and uses of glass types with a view that the user makes a preliminary selection of the type(s) of glass suitable for his application.

Chapter 3 assists the user to select specific type of glass with respect to heat gain (solar incidence), sound insulation, safety & security with respect to thermal breakage, human impact, fire resistance and aesthetics.

Chapter 4 provides tools to workout safe thickness for the specific type of glass selected in previous chapter. Wind pressure varies as per location as well as height of the building. Accordingly appropriate wind pressure is estimated and required thickness of glass is arrived at.

Chapter 5 discusses some of the prevalent glazing systems to fix the glass in buildings.

Chapter 6 gives general guidelines including dos and don'ts for use of glass in buildings. It is followed by references materials and definitions.

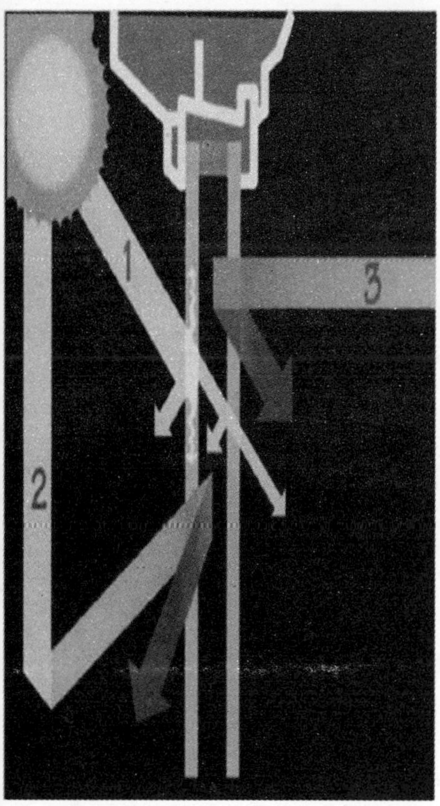

2

Types of Glass, Properties and Uses

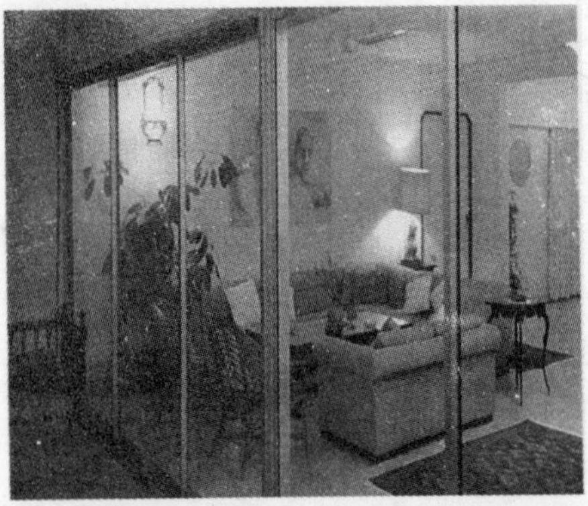

2.0 TYPES OF GLASS, PROPERTIES AND USES

Glass is an inorganic product of fusion that has cooled to a rigid solid without undergoing crystallization. It may be transparent, translucent or shiny depending upon the quantities of its basic constituents such as sand, soda & lime. By varying their quantities & using certain special additives, the properties and characteristics of glass are modulated to obtain a large variety. The important ones used in buildings are discussed in following paragraphs:

2.1 NORMAL (ANNEALED) GLASS

Normal glass is synonymous with flat glass irrespective of the process of manufacture. Float glass has a perfectly flat, brilliant surface, whereas sheet glass has slight distortions. Both are referred as annealed glass and can be processed to obtain many different varieties of glass for use in buildings.

2.1.1 Properties

The properties of normal glass are:

- High light transmission (80 - 90 percent)
- Optical clarity
- Can be processed to produce other glass types such as tempered, laminated and insulating.
- Density (approximate) : 2.42 – 2.52 g/cm^3
- Tensile strength : 40 N/ sq. mm
- Compressive strength : 1000 N/ sq. mm
- Modulus of elasticity : 70 GPa
- Coefficient of linear expansion : 9 x10^{-6} m / mK
- Available thickness : 2 mm - 19 mm
- Normally available sizes up to : 2440 mm x 3660 mm (Bigger sizes can also be made)
- Colour : Clear, Grey, Bronze, Green, Blue and Pink

2.1.2 Use of Normal (Annealed) Glass

Normal glass is used in residences, shopping malls, hotels, restaurants, etc. Other areas where normal glass is used are :

- Windows
- Shelves
- Doors and partitions
- Solar Applications
- Display cases
- Shop fronts
- Solariums
- Greenhouse
- Atriums
- Railings

Metal oxides leading to different tints of glass	
COLORANT	**GLASS COLORS**
Iron	Green, brown, blue
Manganese	Purple
Chromium	Green, yellow, pink
Vanadium	Green, blue, grey
Copper	Blue, green, red
Cobalt	Blue, green pink
Nickel	Yellow, purple
Cadmium	Yellow
Titanium	Purple, brown
Cerium	Yellow
Carbon &	Amber, brown
Selenium	Pink, red
Gold	Red

2.1.3 Types of Normal (Annealed) Glass

2.1.3.1 Clear glass. It is normal annealed glass, though clear & transparent, It does posses some colour tint usually greenish. It is due to presence of some impurities such as iron. Clear glass has very high energy transmission when exposed to sun light. It provides a clear view of the objects across it.

Applications: Used in doors, windows, solar applications, shelves etc. It is also used for further processing to other glass types.

2.1.3.2 Tinted glass: It is normal glass that is coloured by the addition of metal oxides into molten glass. Tinted glass possesses filtering properties that help reduce eyestrain due to dazzle. Its absorption properties help diminish energy transmissions when exposed to sunlight. Tints like green allows more visible light and cut out infra-red radiation.

Variations in the thickness of the glass would yield different performance in terms of light and solar radiation transmission. Although darker shades reduce the amount of heat being transmitted to the interiors, they also reduce the amount of transmitted daylight.

Applications: Used in doors and windows.

2.1.3.3 Patterned, figured or rolled glass: It is a decorative and translucent glass with figures or patterns on one face. In addition to diffusing light and obstructing visibility from the outside, the figures soften the interior lighting. This type of glass is usually more fragile and less convenient to clean.

Applications: Interiors of the buildings, opaque glazing like bathrooms etc.

2.1.3.4 Wired glass: It has wire mesh incorporated during its production. Wired glass is recommended for its fire protection property. In case of fire, the glass cracks but broken pieces tend to remain in position restricting the spread of flame and smoke for some time.

Applications: Used where nominal fire protection is required in windows, doors & partitions.

2.1.3.5 Extra clear glass: Though glass is transparent, it does possess a slight greenish tint due to the presence of some impurities, mainly iron. In display counters and windows, this greenish tint interferes with the true representation of colour and shine. Extra clear glass is a high value glass, free from impurities such as iron. It has high light transmission of more than 92 percent and is free from interference with the true colour & sparkling of objects across it.

Applications: It is used for a sparkling display of expensive materials like jewellery, watches, crystal ware, fine fabrics, art wares, solar applications etc.

2.1.3.6 Ceramic printed glass: Also known as silk-screened glass for its appearance like a silk screen. Certain areas of application make it important to mask a part or whole of glass for privacy or hiding the background or enhancing the look of a product or for purely aesthetical reasons. The size, density and colour would determine the opacity and shading whereas the variety of dots, squares, checks and patterns will give many design combinations to achieve the desired effect.

It is not affected by moisture, oil, soaps, chemicals or detergents and retains its original appearance though out the life of the glass.

Applications: curtain walls, shower installations, glass doors, spandrels and partitions.

2.2 LAMINATED GLASS

Laminated glass is composed of two or more layers of glass with one or more layers of a transparent/ pigmented and specially treated plastic Polyvinyl Butyral [PVB] sandwiched between the glass layers. The glass panes (layers) can be either normal glass or tempered glass. When the glass is broken, fragments tend to adhere to the plastic [PVB] interlayer thereby reducing the risk of injury and helping to resist further damage by weather.

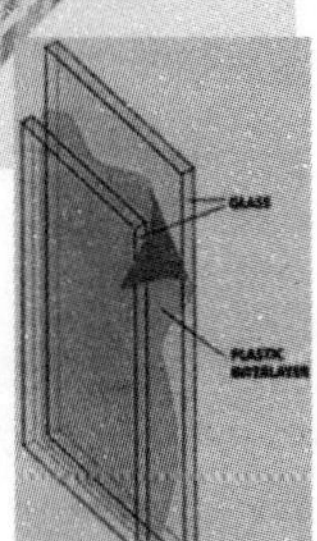

PVB laminated glass

2.2.1 Properties

The properties of laminated glass are:

- Laminated glass does not shatter like ordinary glass. It absorbs impact, resists penetration, and remains intact even if broken, holding glass fragments in place and lowering the risk of injury. Global building standards increasingly specify stricter safety requirements where any breakage could mean a major hazard from falling glass and glass floors.
- Laminated glass resists intrusion because the interlayer continues to safeguard the building even after the glass itself is broken. It cannot be cut from only one sides, so ordinary glasscutters are useless as break-in tools. Laminated glass tends to resist impact. In multi-ply configurations, it can even resist bullets, heavy objects, or small explosions. In most cases, it takes many blows, all in the same spot, to penetrate the glass. The rise in urban crime and terrorism also points to laminated glass as increasingly desirable material.

  Laminated glass is capable to stop flying debris and limit or avoid splintering on opposite side of the impact.

- Laminated glass is an excellent barrier to noise. The sheer damping performance of the plastic interlayer makes laminated glass an effective sound control product. This makes it ideal for airports, hotels, data-processing centers, recording studios, and any building near airports, highways, or train lines.
- Ultraviolet light is the leading cause of deterioration and fading of furnishings, pictures, and fabrics.

Laminated glass screens out 99% of the sun's UV radiation, protecting interior furnishing, displays or merchandise from fading.

- Earthquakes often produce fallout of extremely dangerous shards of broken glass. Also, the heavy winds of tornados and hurricanes easily shatter conventional glass, causing injuries from flying debris and damage to interiors exposed to the devastating weather outside. Laminated glass remains in the frame, maintaining a protective envelope around the home or building to keep weather out and deter glass shards from flying.
- Laminated glass retains its colour and strength for the life of the building and is as easily cleaned as any conventional glass.
- When exposed to heat, laminated glass breaks but stays in place longer. The risk of thermal breakage is avoided only when heat strengthened / tempered laminated glass is used.

- Density (approximate) : 2.42 – 2.52 g/cm^3
- Tensile strength : 32 N/sq. m
- Compressive strength : 1000 N /mm^2
- Modulus of elasticity : 70 GPa
- Coefficient of linear expansion : 9 x 10^{-6} m / mK
- Available thickness : 4.38 mm – 20.76 mm (other thickness can also be made to order)
- Normally available sizes up to : 2000 x 3210mm (Bigger sizes can also be made)

2.2.2 Use of Laminated Glass

Laminated glass is used in office buildings, hotels, restaurants, shopping malls, public walkways, hospitals, libraries, museums, churches, airport terminals, residences

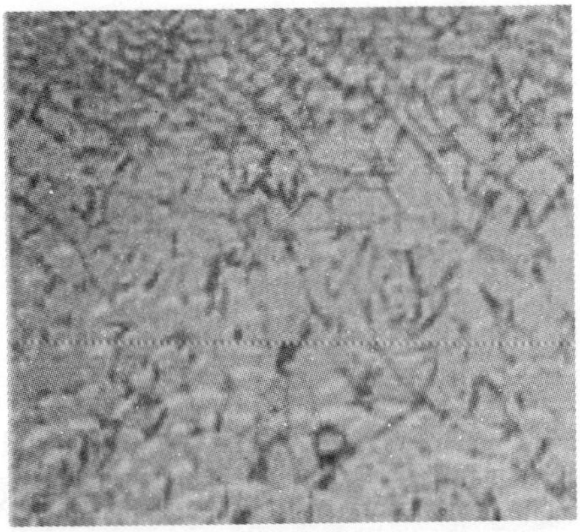

& apartment buildings, noise control applications, embassies, computer centers.

High security places, for example, banks, teller, and drive-through windows, ticket windows, gas stations, currency exchanges, armoured vehicles, jewellery shops and burglar resistant showcases. Other areas where laminated glass is used are:

- Curtain wall glazing
- Sloped glazing
- Skylights
- Glass roofs & floors
- Aquariums
- Animal observatory windows
- Safety glazing for partitions
- Security glazing for banks against bullets/ hand propelled objects.
- Earthquakes, high velocity winds & fire resistance applications.
- Museums
- Acoustic glazing

2.3 TEMPERED OR TOUGHENED GLASS

Tempered glass is an extremely strong glass which is heat treated to a uniform temperature of approximately 650°C and rapidly cooled to induce compressive stresses of 770 kg/m2 to 1462 kg/m2 on the surfaces and edge compression of the order of 680 kg/m2. Tempered glass is not manufactured on float line. It is a separate process.

Tempered or toughened glass gains its added strength from the compressed surfaces. However, if a deep scratch or an impact penetrates the surface, the glass will break into a number of small particles. The heat treatment process for tempered glass requires that all fabrication be completed prior to toughening. Any attempt to cut, drill, grind or sand blast the glass after toughening may result in glass breakage. The heat treatment process does not change the light transmission and solar radiant heat properties of the glass.

2.3.1 Properties

- Tempered Glass is four to five times stronger than normal glass of equivalent thickness. It is mainly used for safety & strength.
- Tempered Glass provides greater thermal strength. It offers increased resistance to both sudden temperature changes and temperature differentials up to 250°C compared with normal glass, which can withstand temperature differentials up to 40°C only.
- Tempered Glass is difficult to break and when on breakage it will break into small, relatively harmless fragments. This substantially reduces the likelihood of injury to people, as there are no jagged edges or sharp corners like normal glass.

Due to the inherent superior features of tempered glass like more strength, ability to withstand sudden impacts and breaking safely into small pieces, it is used as a safety glazing

- Density (approximate) : 2.42 – 2.52 g/cm^3
- Tensile strength : 120 to 200 N/sq. mm
- Compressive strength : 1000 N /sq. mm
- Modulus of elasticity : 70 GPa
- Coefficient of linear expansion : 9 x 10^{-6} m / mK
- Available thickness : 3 mm - 19 mm
- Normally available sizes up to : 2440 mm x 3660mm (Bigger sizes can also be made)

2.3.2 Use of Tempered Glass

It is used in commercial applications where wind, snow or thermal loads exceed the strength capabilities of normal (annealed) glass such as safety glazing for entranceway, railings, partitions or fire knock-out windows. Tempered glass can be used in balustrades, escalator side panels, handrails, shower screens, bathtub enclosures, sliding/swing

doors, squash, racquetball wall, showcases, partitions etc. Other areas where tempered glass should be used are:

- Mainly used for safety and strength
- Curtain walls of high-rise buildings
- Exterior and interior of buildings where strength is important
- Spandrels for walls and decorative panelling
- Door openings, showroom and lobby facades, escalator side plates, and staircase handrails
- Viewing partitions of sports complexes, resorts and airports.

2.4. HEAT STRENGTHENED GLASS

Heat strengthened glass is a type of tempered glass which has been strengthened thermally by inducing a surface compression of 422 to 658 kg/cm^2 as compared to a range of 770 to 1462 kg/cm^2 in case of fully tempered glass. Heat strengthened glass is not manufactured on float line. It is a separate process.

Heat-strengthened glass continues to gain popularity and is often the choice of the design professional for vertical vision spandrel areas and for laminated sloped glazing. It is valued for its mechanical strength, which is twice that of normal annealed glass though half of fully tempered glass. With the exception of strength and breakage characteristics, heat - strengthened glass retains the normal properties of annealed glass, including chemical resistance, hardness, expansion and deflection. Heat-strengthened glass provides necessary resistance to thermal stress associated with high performance glazing materials such as tinted glass and reflective glass. It also provides necessary resistance to heat building up when using spandrel glass. Heat-strengthened glass with its flatter surface also results in the facade having less optical distortions.

2.4.1 Properties

- Density (approximate) : 2.42 – 2.52 g/cm^3
- Tensile strength : 120 to 200 N/sq. mm
- Compressive strength : 1000 N / sq. mm

- Modulus of elasticity : 70 GPa
- Coefficient of linear expansion : 9×10^{-6} m / mK
- Available thickness : 3 mm - 19 mm
- Normally available sizes up to : 2440 mm x 3660mm (Bigger sizes can also be made)

2.4.2 Use of Heat Strengthened Glass

Heat Strengthened glass is suitable for spandrel and vision panels of curtain walls and structural glazing as they safeguards against thermal breakages. It is used for making laminated glass panels for safety combined with strength. It is used in complex glass combinations like double-glazing as one lite of laminated glass for glass floors and roofs.

2.5. HEAT SOAKED TEMPERED GLASS

Tempering process increases the strength of glass many folds, offering a greater resistance to breakage. However, the raw-material used in the manufacture of tempered glass is not free from certain defects including Nickel Sulfide inclusions which put tempered glass at potential risk of spontaneous breakage. The amount and size of these inclusions vary from glass to glass. The process of heat-soaking is used to the transfer of potential on-site breakage to breakage in the factory under a controlled process of accelerating the inversion of the nickel sulfide inclusions to their low temperature phase. In all probability, the glasses that have NiS inclusions will break during this heat soaking process. While heat-soaking does not guarantee that breakage will be completely eliminated in installed tempered glass, it reduces the chances drastically.

2.5.1 Use of Heat Soaked Glass

Heat soaked glass is also used for spandrel and vision panels of curtain walls and structural glazing as they safeguards against thermal breakages. It is ideal for making laminated glass panels for safety combined with strength.

It is also used in complex glass combinations like double-glazing as one lite of laminated glass for poolsides, pool bottoms & aquariums etc. It s an essential requirement to use heat soaked glass for all bolted systems.

2.6 REFLECTIVE GLASS

A metallic coating is applied to one side of the glass in order to significantly increase the amount of reflection by the glass of both the visible and infra-red (light and heat) range of the electromagnetic spectrum.

This metallic coating can be applied to clear or body tinted glass. The reflective glass imparts a mirror like appearance to the exterior of buildings under most daytime conditions. Due to the coating of metal oxides on the glass, they are widely applied as an aesthetic product in buildings for its highly reflective surface and its wide palette of colours. It reduces heat gain and glare from the exterior and allows optimum visible light transmission to the interior. It significantly reduces the air- conditioning load of the buildings. An exceptional property of solar reflective glass is that the coating of metal oxides on the glass can be achieved without affecting the transparency of the glass.

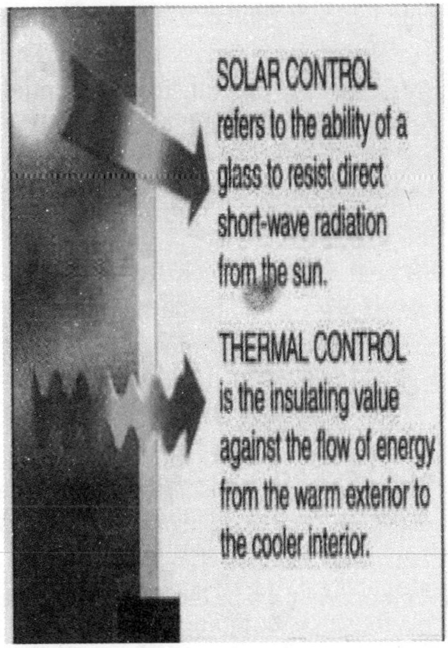

2.6.1 Properties

The properties of reflective glass are:

- Increased aesthetic appeal.
- Gives enormous flexibility in designing the exterior due to availability of number of colours / shades
- Facilitates energy savings through reduction in interior solar heat gain and cost reduction in the cost of heating and cooling systems.
- Improves occupants comfort as interior temperature variations are less and easier to control.
- Varying degrees of light transmittance and varying reflectance.
- Reduces the air-conditioning load of the buildings
- Density : 2.4 – 2.5 g/cm³

- Available thickness : 3 mm - 12 mm
- Normally available sizes up to : 2250 x3210 mm (Bigger size can also be made)

2.6.2 Use of Reflective Glass

Reflective glass is used in office buildings, high-rise buildings,

- Entrance
- Privacy windows
- Decorative walls
- Spandrel glazing
- Vertical and sloped glazing for commercial building applications.
- Solar control applications.
- Building facades.

2.6.3 Types of Reflective Glass

2.6.3.1 Online (Hard) Coated

It is manufactured online during the manufacturing of basic clear or body tinted glass. Gases are injected at high temperature and when glass is still in molten stage (temperature around 650°C) pyrolysis takes place and a layer of metal oxide film is formed in the surface of glass. In online coating the metallic layer becomes part of the glass. Online coating can reflect heat or reduces energy emission plus allows optimum light inside the building. These coatings generally known as Hard Coating or Pyrolytic coatings are very durable and easy in handling and maintenance. Online coating can be solar control coatings or Low e (emissive) coatings.

Solar control coatings are very effective in reducing the direct heat gain or solar heat gain. It reduces the solar factor thus the cooling cost. These coatings are highly recommended for tropical countries like India where the major problem is to cut intense solar radiation.

Low emissive (e) coatings are very effective in cutting the heat flow due to temperature variation between the interior and the exterior. These coating are highly recommended for controlling the heat flow specifically in cold climate when you want to retain the heat inside the building.

Applications: Suitable for further processing to tempered, laminated & insulating glass and also for windows.

2.6.3.2 Offline (Soft) Coating

It is manufactured on a separate processing line where layers of metallic oxides are deposited under vacuum on the surface glass. As it is manufactured in a separate process, it is known as offline coatings. In these coatings the glass is first tempered or heat strengthen and then the coating is applied. Some offline coating can not be used in single glazing. Offline coatings can be solar control or low emissive or some combine the both solar control as well as Low e. Offline coated glasses have high performance as it is selective and allows more visible light and lower infra-red radiations.

2.6.3.3 Temperable Offline Coatings

It is a new generation of coatings manufactured in offline process. These coatings can be further processed like online (hard) coatings. In short they combine the high performance of offline coating with ease of processing of online coatings. They can be also be used in single glazing.

2.7 INSULATING GLASS UNIT (DOUBLE GLAZING)

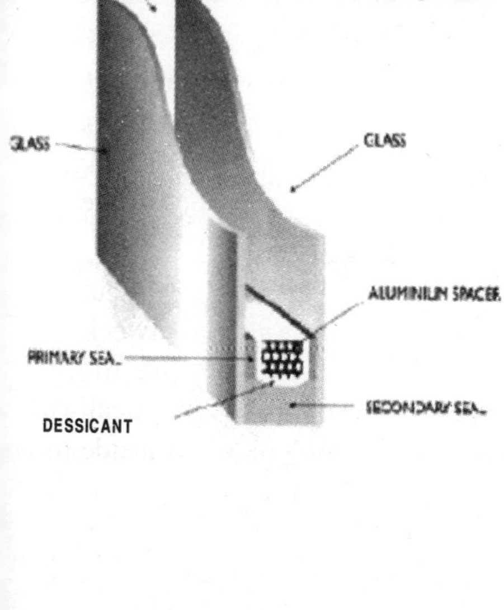

Cross-section of Insulating Glass

The insulating glass is a prefabricated unit made of two or more glass panes, separated by a cavity and edges-hermetically sealed together. This edge seal not only binds the individual sheets of glass together to maintain the mechanical strength of the joint but also protects the cavity between the glasses from outside influences. The moisture in the cavity between the two glasses is controlled by desiccants filled in the perforated spacer. The spacer can be aluminum, composite plastics etc. The spacer ensures the precise distance between the glass panes. The cavity normally filled with dry air but can be also filled with gases such as Argon, Krpton for better thermal performance or hydrogen fluro oxide for better acoustic performance.

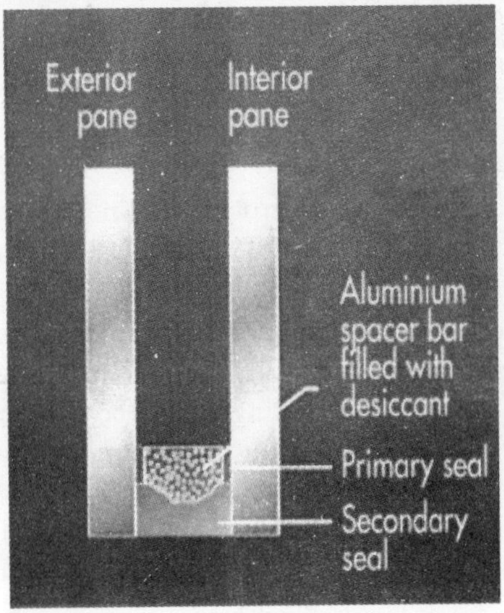

The low heat conductivity of the enclosed dry gas between the glass panes drastically reduces the thermal heat transmission through the glass 2.8 W/sqm-K as compared to 5.73 W/sqm K for normal glass. It also helps is reducing the direct solar energy specifically when the outer pane is a solar control glass.

2.7.1 Properties

- Heat transferred by conduction and convection due to temperature difference between the outside and inside is reduced to nearly half in case of normal glass thus reduces the heat flow / transfer & gain / loss). It is specifically very effective in winters as it saves loss of inner heat.
- The use of heat absorbing or heat reflective glass as outer glass further reduces the load on the cooling system.
- In case of monolithic glass, the temperature difference between the outside and inside of a room may lead to condensation in humid climate. The insulating effect of the air layer makes it difficult for the glass to become cold and is consequently avoids dew condensation.
- Insulating glass can significantly help in reducing the exterior noise pollution if the unit is made up of glass panes of asymmetrical thickness. The amount of sound reduction depends on the combination of the insulating glass. Using one or both panes of laminated or acoustic laminated glasses will drastically reduce sound transmission.
- It offers increased personal comfort and aids energy conservation. Because of its high insulation properties, the lack of cold or warm droughts leads to a pleasant internal environment. Strength to withstand wind load is also increased.
- Normally secondary seal is of silicone if the edges are exposed and of polyshulphide if the edges are framed.

- Available thickness : Custom made
- Normally available sizes up to : Custom made

2.7.2 Use of Insulating Glass

- Office buildings, hospitals, hotels, houses and buildings with exceptionally high heating or cooling requirements.
- Buildings that need the temperature and humidity strictly controlled such as telephone exchanges, laboratories, etc.
- Airport control towers, windows of coaches of trains, and other environments that need regulated atmosphere and prevention of condensation.
- Buildings near highways, railways and airports that need sound insulation property of insulating glass.

2.8 MIRROR

Mirror may be defined as highly reflective glass where the images of objects placed in front of them are formed through its surface by virtue of its reflectivity.

Mirrors are made from float and sheet glass. The reflecting surface is made up of thin coat of metal, generally silver however at times, gold, copper, aluminum or chromium, may also be used if required.

For special mirrors, lead, aluminum, platinum, rhodium, or other metals may be used. The metal film can be semi-transparent or opaque and left unprotected or protected with a coat of shellac, varnish or paint. The silvering metals are deposited on the glass by various means -chemical deposition, amalgams, high-voltage discharge between electrodes in a semi-vacuum, evaporation in a semi-vacuum or painting with organic sulfur compounds of metals. Gold, Green or Bronze mirror is used in Jewellery shops and other showrooms to add aesthetics.

2.8.1 Properties

The properties of mirror are:

- Mirrors have very high reflectivity
- Available thickness up to : 2 mm - 8 mm
- Normally available sizes up to : 2140 mm x 3050 mm

2.8.2 Use of Mirror

Mirrors are used in a number of building types such as offices, hotels, residences, restaurants, airports in

- Wardrobe doors
- Bathrooms
- Dressing rooms
- Display cases
- Decorative walls
- Ceiling and column claddings
- Shelves
- Partitions and Doors
- Jewellery shops.

2.8.3 Types of Mirror

2.8.3.1 Copper Free Mirror

- A coating of Palladium is used instead of copper which is a very durable coating.
- Black spot problem is eliminated.
- It has longer life as compared to normal mirror.

2.8.3.2 Copper Free Lead Free Mirror

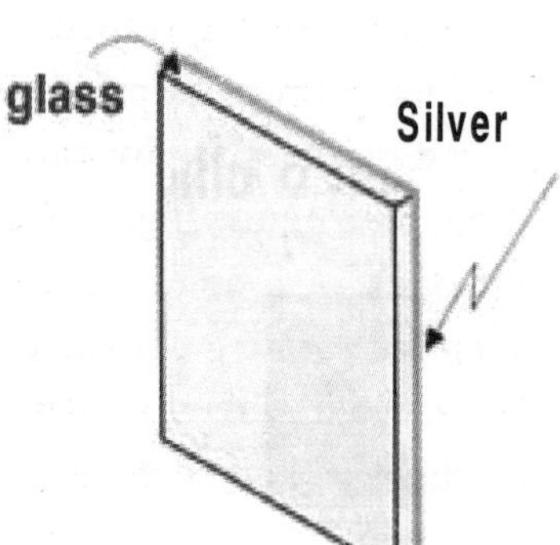

- A coating of Palladium is used instead of copper and the protective paint used is free from lead hence it is more environment friendly.
- Improves optical clarity and has significantly longer life compared to normal mirrors

2.8.3.3 UV Coated Mirror

- UV treated polymer coating is applied over paint coat by a roller.

- The coating provides,
  - o Better scratch resistance.
  - o Acts as a second protective coat on the paint.

2.8.3.4 Safety Mirror

- It is a new generation of mirrors with a special film applied on it gives the glass a safety performance and classification.

General properties and common uses of normal glass, laminated glass, tempered glass, reflective glass, insulating glass & mirror in different buildings have been described in this chapter with a view that the reader makes a preliminary selection of the type of glass appropriate for his application.

3

Selection of Specific Glass

3.0 SELECION OF SPECIFIC GLASS

3.1 GENERAL

In the previous section, a general selection of the types of glass has been discussed. The present section discusses the selection of specific types of glass for particular applications. Considering the ever-increasing options of variety of glass products, it is essential to consider the factors that influence the choice of a particular type of glass to get the desired results. Selection of glass for use in buildings on scientific lines depends upon the following parameters:

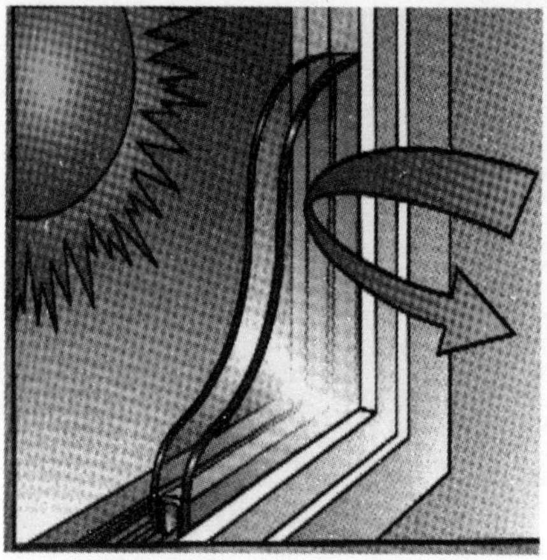

- Heat Gain (solar incidence)
- Sound insulation
- Safety & Security:
  - o Thermal Breakage
  - o Human Impact
  - o Security (burglar proof, Bullet Proof, Explosion Proof etc.)
  - o Fire Resistance
- Strength
- Aesthetics

3.2 HEAT GAIN

Different types of glass treat the incident solar energy in different ways. It involves:

- Heat transfer mechanism
- Solar energy spectrum
- Factors determining the performance of glass
- Total heat gain
- Selectivity in tropical countries

3.2.1 Heat Transfer Mechanism

Heat transfer happens in three ways - conduction, convection and radiation:

Conduction: Flow of heat from hotter part to cooler part in a solid.

Convection: Heat transfer due to movement of liquid or gas. It does not happen in vacuum.

Radiation: Heat transfer due to heat emitted by hot bodies. Heat emitted by sun is know as solar radiation and by bodies on earth are know as thermal radiation. There is no medium required for heat transfer by radiation.

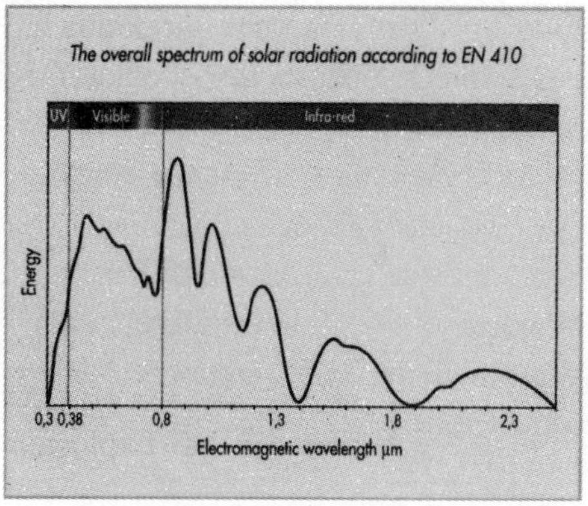

3.2.2 Solar Energy Spectrum

Energy coming from solar radiation at terrestrial level is sub-divided into:

Ultra Violet Rays: 3%

Visible rays: 42%

Infrared radiation: 55%

3.2.3 Factors Determining the Performance of Glass

Factors, which determine the performance of glass are defined as follows:

1. Visible light transmittance (%): The percentage of light transmitted in the visible spectrum, from 380 to 780 nanometers.
2. Visible light reflection (%): The percentage of visible light reflected
3. Ultra Violet Rays (%): Percentage of amount of ultra violet rays transmitted.
4. Solar Energy Transmission (%): Percentage of incident solar energy transmitted by glass.
5. Solar Energy Reflection (%): Percentage of incident solar energy reflected by glass.
6. Solar Energy Absorption (%): Percentage of incident solar energy absorbed
7. Solar Factor (SF): Is the amount of energy entering the building, expressed as a percentage of the incident solar energy. Lower the solar factor, lower the solar energy heat gain.

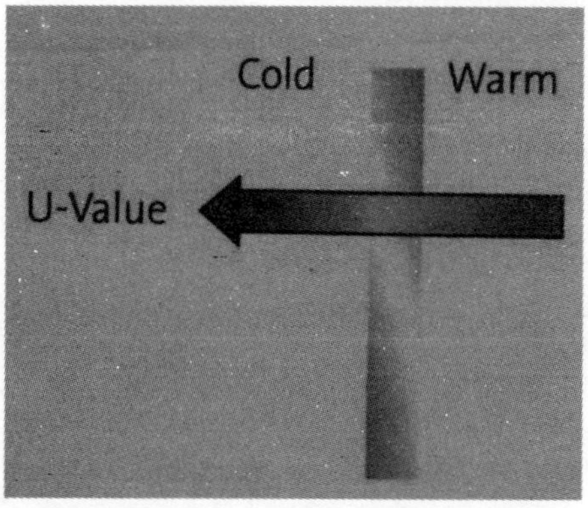

8. Shading Coefficient (SC): Ratio of solar factor of glass to solar factor of 3mm clear glass (0.87)
9. U Value (W/sqm K): Rate of heat transfer through 1sqm of glass for a temperature differential of 1° K between the interior and exterior. The thermal insulation of glass is assessed by its U-value. Lower the U-value, the lower the heat transfer (loss or gain) through glass.

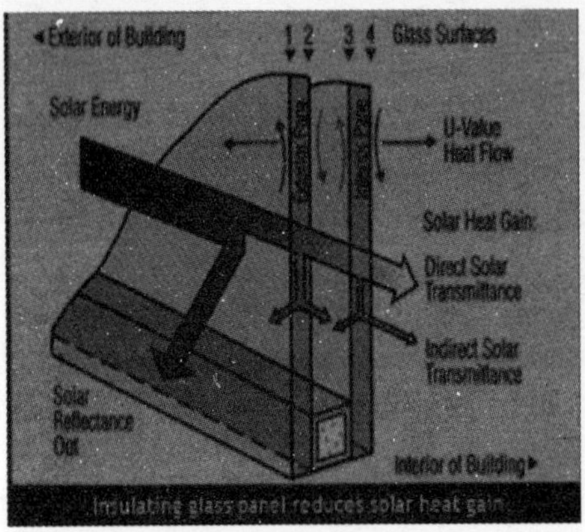

Insulating glass panel reduces solar heat gain

3.2.4 Total Heat Gain

Total heat gain through glass is important especially in tropical countries as it has direct implication on cooling cost. Lower the heat gain means lower the cooling cost. It is the heat transferred due to temperature difference between inside and outside plus the heat gain due to incident solar energy. In tropical countries the contribution of heat gain due to incident solar energy is over 80% of total energy hence it is very important and essential to minimize amount of heat gain due to direct solar energy which is controlled by solar factor (SF) of glass.

3.2.5 Selectivity in Tropical Countries

Light Transmission needs to be optimized so as to avoid glare and minimize artificial illumination. It mainly depends on use of buildings. Retail / shopping areas may require maximizing light transmission due to higher visibility whereas software parks may require minimizing light transmission to reduce glare.

Glass needs to transmit true colour, which means the colour of light transmitted through the glass should be close to that of light through the clear glass. This is essential to avoid false lighting and giving a perfect working atmosphere.

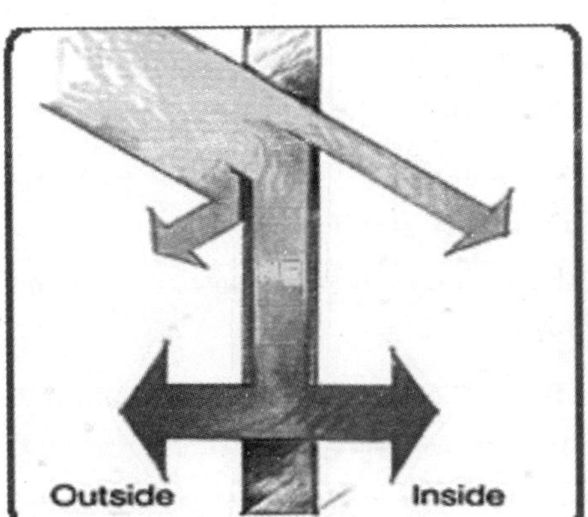

Solar control by body-tinted glass is due to its absorption property of heat. They are also know as heat absorbing glasses and provide medium level of solar control.

Solar control by reflective glasses is due reflection of solar energy and produce very good solar control. It is most effective on sun facing wall. The optimum position of solar control coating is on external surface of the building. All hard

coatings are durable for this application. Soft and medium solar coating cannot be positioned on external surfaces.

Body tinted glasses with reflective coatings achieve solar control by absorbing as well as reflecting solar heat energy. They combine both the functions. Reflective glass is best for reducing peak air-conditioning load requirements. Double glazed or insulating glass unit helps in further reducing the operating costs. Saving in capital and operating costs of AC is best achieved by using a double glazing with solar control coated glass on the outside with clear or low e glass on the inside.

Any solar control product, however selective in its absorption or reflection of solar radiation, will reduce the light transmission. Though ultraviolet constitutes only 3% of the solar spectrum, it is the major cause of fading curtains, carpets and upholstery inside a building. Laminated glass cuts down almost 99% of all ultraviolet radiation and drastically reduces fading.

Indicative values of the factors determining the performance of glass (photometric properties) with respect to the glass type covered in the study are annexed in table 3.1 to 3.9.

Table 3.1 : Normal (Annealed) Glass (Clear Glass - Photometric Properties)

Thickness (mm)	U value (Watts / sq.m. k)	UV Transmittance %	Visible Light			Solar Energy		
			Transmittance	Reflectance		Reflectance (Out)	Transmit-tance	Shading Coefficient
				Indoor	Outdoor			
2	3.53	77	90	8.1	8.1	7.6	86	1.00
3	3.53	72	90	8.0	8.0	7.4	83	0.99
4	3.47	67	89	7.9	8.0	7.2	80	0.97
5	3.44	64	88	7.9	7.9	7.0	77	0.95
6	3.44	61	87	7.8	7.8	6.9	75	0.93
8	3.37	56	86	7.7	7.7	6.6	70	0.89
10	3.34	52	85	7.6	7.6	6.4	66	0.86
12	3.28	49	83	7.5	7.5	6.2	62	0.83
15	3.22	44	81	7.3	7.4	6.9	57	0.80
19	3.15	39	78	7.1	7.2	5.7	51	0.75

Table 3.2 : Normal (Annealed) Glass (Bronze - Photometric Properties)

Thick-ness (mm)	U value (Watts / sq.m. k)	UV Transmit-tance %	Visible Light			Solar Energy		
			Transmit-tance	Reflectance		Reflectance (Out)	Transmit-tance	Shading Coefficient
				Indoor	Outdoor			
3	3.47	36	66	6.3	6.5	6.3	64	0.85
4	3.47	28	58	5.9	6.1	5.9	57	0.79
5	3.44	22	52	5.5	5.7	5.6	51	0.74
6	3.44	18	46	5.3	5.5	5.3	45	0.70
8	3.37	12	37	4.9	5.1	5.0	36	0.63
10	3.34	08	29	4.6	4.9	4.8	29	0.57
12	3.28	5	23	4.5	4.8	4.7	23	0.53

Table 3.3 : Normal (Annealed) Glass (Grey - Photometric Properties)

Thick-ness (mm)	U value (Watts / sq.m. k)	UV Transmit-tance %	Visible Light			Solar Energy		
			Transmit-tance	Reflectance		Reflectance (Out)	Transmit-tance	Shading Coefficient
				Indoor	Outdoor			
3	3.50	37	62	6.2	6.4	6.3	63	0.83
4	3.47	29	55	5.8	6.0	5.9	55	0.77
5	3.44	23	48	5.4	5.7	5.6	49	0.72
6	3.44	19	42	5.1	5.4	5.4	43	0.68
8	3.37	13	32	4.8	5.1	5.1	34	0.61
10	3.34	09	25	4.6	4.9	4.9	26	0.55
12	3.28	06	19	4.4	4.8	4.7	21	0.51

Table 3.4 : Laminated Glass (Photometric Properties)

Type of Glass	Visible Light Transmittance [%]	UV Transmittance [%]	Shading Coefficient	Ratio of VLT/UVT	U Value [Watts/ sq m. ^{0}K]
Laminated Glass for 6 mm thickness					
Clear	89	77	0.92	1.16	1.05
Blue Green	73	68	0.84	1.07	
Translucent White	65	58	0.76	1.12	
1. Light bronze	52	54	0.72	0.96	
2. Medium bronze	28	34	0.56	0.82	
Grey	44	47	0.67	0.94	
Blue Grey	28	39	0.60	0.72	
Laminated Insulating Gla					

3 mm thick					
Clear	79	62	0.78	1.27	0.50
Blue Green	66	56	0.72	1.18	
Bronze [light]	46	43	0.60	1.07	
Neutral Brown [light]	49	44	0.63	1.11	
Laminated Insulating Glass 6 mm thick					
Grey	39	36	0.54	1.08	0.46
Translucent White	58	45	0.63	1.29	

Table 3.5 : Mirror (Photometric Properties)

Thick ness (mm)	U value (Watts / sq.m. k)	UV Transmit tance %	Visible Light			Solar Energy		
			Transmit -tance	Reflectance		Reflectance (Out)	Transmitt -ance	Shading Coefficient
				Indoor	Outdoor			
6.3	3.53	0	0	93	0	0	0	0.35
5.9	3.50	0	0	92	0	0	0	0.36
5.6	3.47	0	0	90	0	0	0	0.36
5.4	3.47	0	0	89	0	0	0	0.36
5.1	3.44	0	0	87	0	0	0	0.36

Table 3.6 : Reflective Single Glazing (6 mm Thickness - Photometric Properties)

Description	Coating	U value (Watts / sq.m. k)	Visible Light			Solar Energy		
			Trans mittance	Reflectance		Reflec tance (Out)	Transmit tance	Shading Coefficient
				Indoor	Outdoo			
Silver	SS - 8	2.84	8	37	41	34	7	0.23
	SS - 14	2.96	14	35	31	27	11	0.30
	SS - 20	3.12	20	32	25	22	16	0.37
Sterling	CS - 14	2.90	14	37	33	27	12	0.30
	CS - 20	3.00	20	32	27	22	17	0.37
Silver-Blue	TS - 20	3.00	20	33	23	22	14	0.35
Blue	TS - 30	3.15	30	29	17	15	24	0.46
Deep-Blue	TS - 40	3.37	40	24	10	10	32	0.55
Green	S - 8	2.84	6	37	34	19	4	0.25
	SS -14	2.96	11	35	27	16	6	0.29
	SS -20	3.12	16	32	20	13	9	0.34
Blue Green	TS -20	3.00	17	33	18	12	9	0.34
	TS -30	3.15	26	29	14	9	15	0.40
Bronze	SS - 8	2.84	5	37	17	15	4	0.26
	SS -14	2.96	8	35	15	13	6	0.30
	SS -20	3.12	12	32	12	11	10	0.35
Grey	SS - 8	2.84	4	37	13	15	4	0.26
	SS -14	2.96	7	35	10	12	6	0.31
	SS -20	3.12	10	32	9	10	9	0.35
Silver Blue Grey	TS -20	3.00	10	33	9	11	8	0.33
Sapphire Blue-Grey	TS -30	3.15	15	28	7	8	14	0.40

Table 3.7 : Reflective Double Glazing (25 mm - Photometric Properties)

Description	Coating	U value (Watts / sq.m. k)	Visible Light			Solar Energy		
			Transmittance	Reflectance		Reflectance (Out)	Transmittance	Shading Coefficient
				Indoor	Outdoor			
Silver	SS - 8	1.26	7	38	41	34	6	0.16
	SS - 14	1.32	13	37	31	27	9	0.21
	SS - 20	1.39	18	34	25	22	13	0.27
Sterling	CS - 14	1.29	13	38	33	28	10	0.22
	CS - 20	1.32	18	34	27	23	15	0.28
Silver-Blue	TS - 20	1.39	18	35	23	22	12	0.25
Blue	TS - 30	1.42	27	31	17	15	20	0.35
Deep-Blue	TS - 40	1.48	36	27	11	10	20	0.35
Green	SS - 8	1.26	5	38	34	19	3	0.15
	SS - 14	1.32	10	37	27	16	5	0.18
	SS - 20	1.39	15	34	20	13	8	0.22
Blue Green	TS - 20	1.39	16	34	18	12	8	0.23
	TS - 30	1.42	23	31	14	9	13	0.28
Bronze	SS - 8	1.26	4	38	17	15	3	0.16
	SS - 14	1.32	7	37	15	13	5	0.19
	SS - 20	1.39	11	34	12	11	8	0.23
Grey	SS - 8	1.26	4	38	13	15	3	0.16
	SS - 14	1.32	6	37	10	11	5	0.19
	SS - 20	1.39	9	34	9	10	8	0.23
Silver Blue Grey	TS - 20	1.39	9	34	9	11	7	0.22
Sapphire Blue-Grey	TS - 30	1.42	13	31	7	8	12	0.28

Table 3.8 : Reflective Glass (6 mm Thickness - Photometric Properties)

Description	Coating	U value (Watts / sq.m. k)	Visible Light			Solar Energy		
			Transmittance	Reflectance		Reflectance (Out)	Transmittance	Shading Coefficient
				Indoor	Outdoor			
Harvest Bronze	CB – 8	2.84	5	55	13	13	4	0.27
	CB – 14	3.03	8	50	10	10	8	0.33
	CB – 20	3.22	12	44	8	8	11	0.38
Misty Bronze	CP – 8	2.87	5	53	14	13	4	0.27
	CP - 14	3.12	8	47	11	10	8	0.34
	CP - 20	3.22	12	41	10	9	12	0.38
Charren Gray	CB – 8	2.84	4	55	10	12	4	0.27
	CB - 14	3.03	7	50	8	10	8	0.33
	CB - 20	3.22	10	44	7	8	11	0.38
Misty Gray	CP – 8	2.87	4	53	11	13	4	0.28
	CP - 14	3.12	7	47	9	10	8	0.34
	CP - 20	3.22	10	41	8	8	12	0.38
Royal Indigo Gray	CR – 8	2.81	4	55	9	11	5	0.28
	CR - 14	3.00	7	50	8	8	8	0.33
	CR - 20	3.19	10	44	6	7	11	0.38
Platinum	SP - 13	2.93	13	43	29	23	12	0.31
	SP - 18	3.06	18	38	24	21	16	0.36
	SP - 22	3.12	22	36	19	17	20	0.41
	SP - 33	3.25	33	27	11	10	29	0.52
Serenity Green	SP - 13	3.06	11	43	23	13	7	0.31
	SP - 18	3.12	15	38	19	12	10	0.34
	SP - 22	3.25	19	36	15	10	12	0.37
	SP - 33	2.93	28	27	9	7	18	0.44
Satin Bronze	SP - 13	3.06	8	43	10	11	7	0.31
	SP - 18	3.12	11	38	9	11	10	0.34
	SP - 22	3.25	13	35	8	9	12	0.37
	SP - 33	2.93	20	27	6	6	18	0.44
Starlite Gray	SP - 13	3.06	7	43	10	11	7	0.31
	SP - 18	3.12	9	38	9	10	10	0.34
	SP - 22	3.25	11	35	8	9	12	0.37
	SP - 33	2.84	16	27	6	6	18	0.44

Non-Coated Glass

Clear Normal	3.56	89	8	8	7	81	0.96
Clear Laminated	3.37	86	7	7	6	67	0.86
Green Tint	3.56	76	7	7	6	48	0.71
Bronze	3.56	53	6	6	5	48	0.70
Grey	3.56	43	5	5	5	47	0.70

Table 3. 9 : Insulating Glass (Photometric Properties)

Description	Nominal Thickness, mm	U value 6 mm air space	U value 12.9 mm air space	Visible Light Transmission %	Solar Radiant Heat Reflection %	Solar Radiant Heat Absorption %	Direct Solar Transmission %	Total Solar Transmission %	Shading Coefficient No Shade	Shading Coefficient Venetian Blinds – Light	Shading Coefficient Venetian Blinds – Medium	Shading Coefficient Drapes – Light	Shading Coefficient Drapes – Medium
Clear Normal & Monolithic	6	1.13	1.13	87	8	12	80	84	0.95	0.55	0.64	0.56	0.61
Clear Normal & Clear Normal	6	0.64	0.56	76	13	23	64	73	0.83	0.51	0.57	0.48	0.53
Clear Normal & Bronze Normal	6	0.64	0.56	44	7	58	35	46	0.52	0.36	0.39	0.38	0.41
Clear Normal & Grey Normal	6	0.64	0.56	37	7	58	35	46	0.52	0.36	0.39	0.38	0.41
Clear Normal & Green Normal	6	0.64	0.56	43	13	42	45	55	0.62	0.43	0.46	0.46	0.49

3.3 SOUND INSULATION

Sound insulation is desirable for peaceful and more productive environment in buildings. Sound intensity is the amount of energy in a sound wave and is proportional to sound pressure, expressed in decibels (dB). Under typical conditions, an individual with normal hearing cannot detect a change in sound pressure of 1-2 dB, while a pressure of 3dB is barely noticeable if it is sustained with no time lapse in between. A difference of 5dB is clearly detected. Sound reduction level achieved by any type of glass will be different at different frequencies. Sound arises from molecules vibrating in a gas, liquid or solid. The number of vibrations or sound waves emitted per second is known as the frequency and is expressed in Hertz (Hz). The human ear is sensitive to sounds in the frequency range 16Hz to 20,000 Hz. Architectural acoustics generally concerns itself with the 50 HZ to 5000 Hz range,

The sound insulation from a partition is approximately given by

$$R_{AV} = 10 + 14.5 \log_{10} m$$

Where R_{AV} = Average sound reduction in dB

m = mass / unit area in kg / m^2.

The greater the mass, the larger the insulation provided. A 23 cm brick wall has a mass of 415 kg/m^2 and gives an insulation of about 50 dB. A 6mm normal glass has a mass of 130 kg/m^2 and gives an insulation of about 40 dB.

Optimum performance can be achieved from a partition when it provides good acoustic insulation at the frequencies where the noise is at its greatest. Good acoustic insulation against specific types of noise can also be achieved by modifying the type and composition of the glazing. For this reason a weighted sound reduction index, Rw (C;Ctr) has been created. The correction Ctr or C is used depending upon if the source of the noise in question is road traffic or outside background noise respectively. Both corrections are generally negative and are deducted from the Rw to determine the noise reduction properties of a building element. They are provided by test laboratories and appear alongside the value for Rw.

Which values to use where:		
Source	C	Ctr
Childrens Playing	X	
Domestic activities	X	
Disco Music		X
Rapid road traffic (>80km/hr)	X	
Slow Road Traffic		X
Medium to high Speed traffic	X	
Slow Rail Traffic		X
Air Traffic Short Distance	X	
Air Traffic Long Distance		X
Air Traffic with Propellers		X
Industry: Medium to high frequency	X	
Industry: Medium to low frequency		X

For example according to EN 717-1, Rw (C;Ctr) = 37 (-4;-9) means that the sound insulation value for a façade is 37 dB and is reduced by 9 dB for traffic noise. The table 3.12 shows the Rw (C;Ctr) values of various glass types currently available in the market.

When a partition consists of more than one material e.g. a glass window fixed in a brick wall, the sound insulation of the partition is regulated by proportional areas of each material. Even minor holes or leakages drastically alter the acoustical performance. The overall sound performance also depends on the total framing and other construction materials. A hole of only 1/100 of the total area would reduce the 50 dB insulation to about 20 dB.

Sound reduction is achieved even by use of single glazing. As the thickness of the glass increases the sound insulation also increases. PVB

Laminated glass enhances its sound insulation property due to addition of one more medium plus effect of the resonance is lower. Specialized acoustic PVB laminated glasses further enhance the performances by bridging the critical frequency thus nullifying the effect of resonance.

Insulating units constructed with equal panes typically exhibit a resonant frequency during which both panes vibrate simultaneously resulting in a significant loss in performance. It is, therefore recommended that glass panes of different thickness be used. Thickness difference between inner and outer glass is recommended to be greater than or equal to 30 percent.

Acoustic performance by insulating glass unit can be further enhanced by using one or both the panes made up of PVB laminated glasses. When both the glasses are laminated then using glass of different thicknesses can further enhance performance. For further better performance acoustic PVB instead of normal PVB should be used. Acoustic PVB has a sound-damping layer sandwiched between the two PVB.

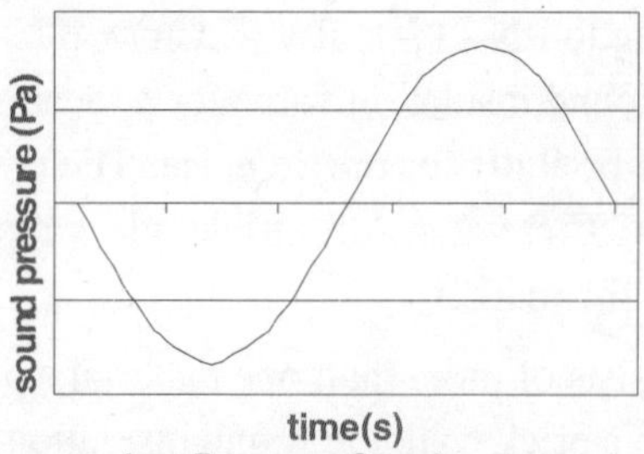

Low Frequency Sound

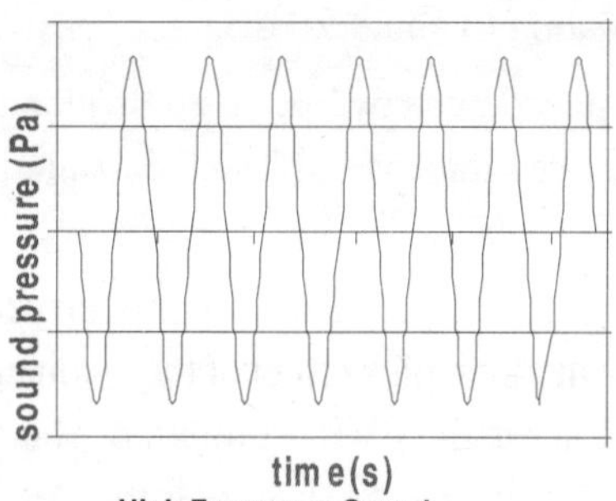

High Frequency Sound

Insulating units assembled with intermediate airspace filled with special gases i.e. sulphur hexa - fluoride (SF6) are known for sound reduction to significant levels.

Table 3.10 gives the noise levels of important noise sources. As the noise level reduces with the distance, these values may be modified according to the distance of the building from the source. The modified noise level may be calculated as follows:

Modified Noise level = Noise level at source - 20 $\log_{10} r$ – 10.9 where r is the distance of the building from the noise source.

Ambient noise level and noise reduction index, R_w being known, the inside noise level is regulated by

Inside noise level = Ambient noise level – R_w + $\log_{10}$ S. where S is the area of glass / partition through which the sound is transmitted in m^2.

The inside noise level is further governed by the acoustic absorption in the room.

Table 3.11 gives the acceptable indoor noise levels in various buildings. Depending on the ambient noise level & acceptable noise level, appropriate glazing may be worked out from table 3.12.

Table: 3.10: Acceptable Indoor Noise Levels for Various Buildings (as per BS 8233):

Location	Recommended max noise level
Dwellings :	
Bedrooms	30 - 40
Living rooms	40 - 45
Offices :	
Private offices and small conference rooms	40 - 45
Large offices	45 - 50
Educational :	
Classrooms and small lecture rooms	40
Large classrooms, large lecture rooms	
and language laboratories	35
Music and drama areas	30
Health and Welfare:	
General wards	55
Small consulting rooms	50
Diagnosis rooms	45

Table 3.11: Sound Levels of Some Noisy Sources

Type of noise	dB	Conversation	Effect on hearing
Space rocket	175	impossible	irreversible damage
Jet plane	135		
Noisy workshop	**120**	**threshold of pain**	
Pneumatic drill	100	shouting	
Underground train	90		damage may occur
Street traffic	80	difficult	
Television	70	with raised voice	
Large store	60		
Calm apartment	50		
Quiet room	40	normal voice	audible noise
Country residence	30		
Rustling leaves	20	hushed voices	
Recording studio	10		
Total silence	**0**	**threshold of hearing**	

Table 3.12: Sound Reduction through Various Glass Combinations in dB.

GLASS TYPE	CONFIGURATION	Rw (C;Ctr) (dB)
Monolithic Glazing		
6 mm Float	6 mm	31 (-1;-2)
8 mm Float	8 mm	32 (-1;-2)
10 mm Float	10 mm	33 (-1;-2)
Laminated Glass with normal PVB		
6.38 mm PVB Laminate	3mm/0.38mm/3mm	33 (-1; -4)
8.38 mm PVB Laminate	4mm/0.38mm/4mm	34 (-1;-4)
10.38 mm PVB Laminate	5mm/0.38mm/5mm	36 (-1;-4)
Laminated Glass with Acoustic PVB		
6.38 mm Acoustic PVB Laminate	3mm/0.38mm/3mm	36 (-1;-3)
8.38 mm Acoustic PVB Laminate	4mm/0.38mm/4mm	37 (-1;-2)
10.38 mm Acoustic PVB Laminate	5mm/0.38mm/5mm	38 (-1;-2)
Normal Double Glazing		
4mm Float – 12mm air gap – 4mm Float	4 – 12 – 4mm	30(0;-3)
6mm Float – 12mm air gap – 6mm Float	6 – 12 – 6mm	33(0;-3)
Asymmetrical Double Glazing		
6mm Float – 12mm air gap – 4mm Float	6 – 12 – 4mm	33 (-1;-4)
8mm Float – 12mm air gap – 6mm Float	8 – 12 – 6mm	35 (-2;-5)
8mm Float – 12mm air gap – 10mm Float	8 – 12 – 10mm	37 (-1;-3)
Double glazing with one pane laminated		
6mm Float – 12mm air gap – 8.38mm Float	6 – 12 – 8.38mm	37 (-1;-4)
8mm Float – 12mm air gap – 8.38mm Float	8 – 12 – 8.38mm	38 (-2;-4)
Double glazing with one pane acoustic PVB laminated		
6mm Float – 12mm air gap – 8.38mm Float	6 – 12 – 8.38mm	38 (-1;-5)
8mm Float – 12mm air gap – 8.38mm Float	8 – 12 – 8.38mm	40 (-2;-5)
Double glazing with two pane acoustic PVB laminated		
8.38mmFloat – 2mm air gap–10.38mm Float	8.38 – 12 – 10.38mm	43 (-2;-6)
8.76mm Float – 20mm air gap–10.76mm Float	8.78 – 12 – 10.78mm	47 (-2;-7)

3.4 SAFETY & SECURITY

Glazing areas are increasing with modernization leading to question of safety of glazing against human impact, injury to human and security against intrusion /bullets and explosion. This section briefly covers each one of them.

- o Thermal Breakage
- o Fire resistance
- o Safety: Human Impact
- o Security:
  - Burglar proof
  - Bullet Proof
  - Explosion Proof
  - Earthquakes
  - Hurricanes

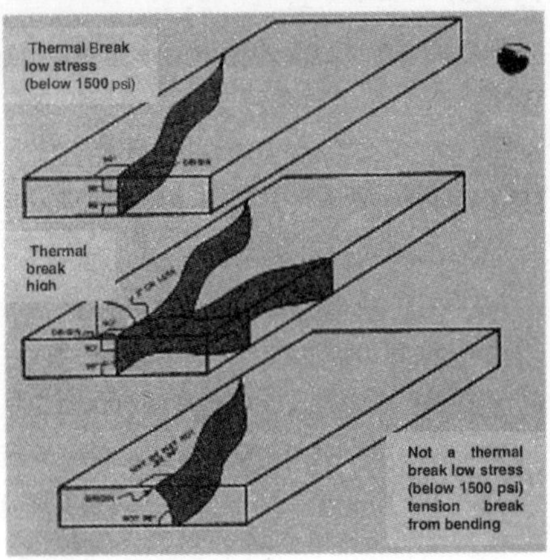

3.4.1 Thermal Breakage

Thermal breakage is generated by exposure of a portion of the glass to the heat of the sun while the other portion remains cool. Resistance of the glass to thermal stress depends on the following:

- Heat absorption properties of the glass range from 8% (low risk of thermal breakage) to 30-40% (medium risk of thermal breakage) to 80-85% (high risk of thermal breakage)
- Heat transfer properties of the cavities of insulating units.
- Location of the building
- Exposure/orientation of the glass
- Fixing details of the glass (frame type and design)
- Presence of blinds or heavy curtains close to glass etc.
- Colour of blinds / curtain

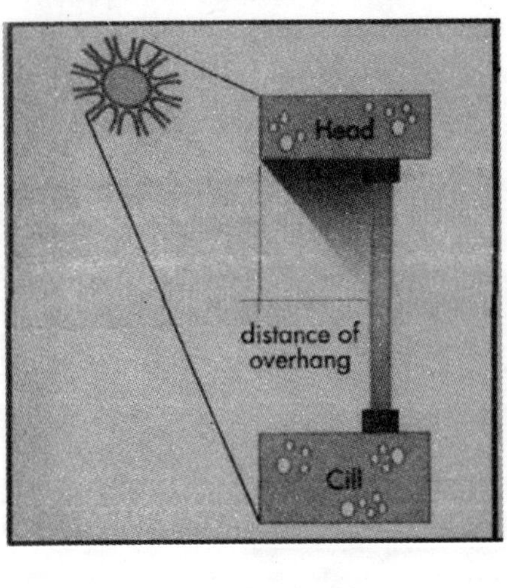

Thermal crack always starts at 90° to the edges and further develop normally in a lazy / zigzag crack pattern. Glass that is not heat-treated (heat strengthened or tempered) is susceptible to thermal breakage. Tempered glass and heat-strengthened glass minimize possibilities

of thermal breakages while laminated and double-glazed glass without tempering or heat strengthening have the chances of thermal breakages. Therefore, if the risk of thermal breakage is high, heat strengthened glass should be used. Thermal cracks in glasses used in annealed condition can be minimized by having clean cuts, grinding the edges, non-chipped off edges and minimizing rebate in the frames plus ensuring gaskets in place.

3.4.2 Fire Resistance

Glass does not burn as it is not combustible but normal glass does not prevent spread of fire as it easily cracks when subjected to heat. It also does not have any natural heat shielding capability. Fire resistance may be achieved by the following:

- Appropriate framing of the glass.
- Preventing the cracked glass from falling apart e.g. wired glass, laminated glass with special interlayer.
- Using low thermal expansion glass such as borosilicate glass or ceramic.
- Making glass more resistant to cracking by toughening or heat strengthening.

For emergency exit passages in large public or commercial building, only approved fire rated glasses should be used. For selecting the correct fire rated glass, the requirements for the glazing system have to be considered as under:

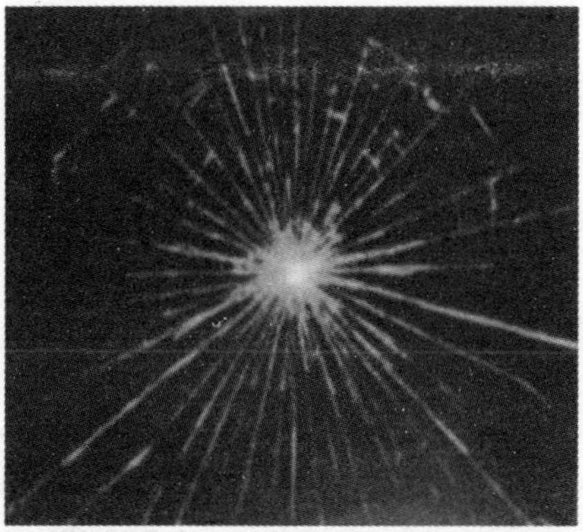

- Integrity only: The glass will not collapse for a set time (usually 30 or 60 minutes).
- Integrity plus Insulation: Added ability to restrict the surface temperature on non-fireside to remain below an average of 140° C.
- Integrity plus Radiation Control/Partial Insulation: Additional capability to restrict the radiation on the non-fireside to less than 15Watts per sq m.

Wired glass is most common type of glass for 30 minutes rating for integrity. For Insulation, double-glazed wired glass will provide integrity for up to 60 minutes and

Insulation for up to 30 minutes. It is not only glass, but the total glazing system has to be fire rated. The provision of National Building Code for fire protection should be followed.

3.4.3 Safety: Human Impact

The concept of safety is related to human safety i.e. any person accidentally coming in contact with glass should not be injured. This does not presuppose that the glass will not be broken under impact conditions, but rather that it will not be broken under the most likely forms of impact or, when broken, the likelihood of cutting and piercing injuries will be minimized. Recommends for use of Grade 'A', & 'B' safety glazing materials. Grade 'A' being the stronger safety glass is followed by Grade 'B'. Tempered & laminated safety glasses are generally Grade 'A' while normal (annealed) & safety wired glass are Grade 'B' safety materials

- Glass in doors shall be Grade 'A' safety glazing materials in accordance with table 3.13 except that:
    - o In fully framed panels, normal glass up to a maximum area of 0.5 m^2 may be used.
    - o Unframed doors shall be glazed with toughened safety glass with a standard nominal thickness of not less than 10 mm

All framed glass in side panels with their nearest vertical sight line less than 300 mm from the nearest edge of the doorway opening shall be of Grade 'A' safety glazing material in accordance with table 3.13 except that:

- o In non-residential buildings, normal glass in accordance with table 3.14 may be used in any of the following cases:
    - i) where the clear opening width of the glass is not wider than 500 mm at any part.
    - ii) where the lowest part of the glass is 500 mm or more above the highest abutting finished floor level.
    - iii) where the clear opening height of each of the glass panels is no greater than 1000 mm at any part.

iv) where the side panel is provided with a chair rail to protect each face of the glass and located with its upper edge not less than 700 mm or its bottom edge not more than 1000 mm above the highest abutting finished floor level.

- In non-residential buildings, normal glass in accordance with table 3.16 may be used provided that the clear opening width of the glass is greater than 2 m and either-

  i) A stall board not less than 300 mm above the highest abutting finished floor level is provided or

  ii) ordinary normal glass not less than 10 mm standard nominal thickness is used.

- In residential buildings, where the lowest visible sight line is 1200 mm or greater above the highest abutting finished floor level, normal glass in accordance with table 3.16 may be used.

- Unframed glass in side panels without exposed edges shall be used with Grade 'A' safety glazing material in accordance Table 3.13 or Table 3.17.

- Louvres in side panels with any edge exposed shall be glazed with a Grade 'A' tempered safety glazing material in accordance of table 3.13 or 3.17 except that a standard nominal thickness of not less than 5mm for louvers shall be used up to 225 mm in blade width. For blade widths in excess of this, a minimum of 10 mm glass shall bc uscd.

- A glazed panel so located in a building that it is capable of being mistaken for an unobstructed opening shall comply, as appropriate, with the following requirements:

  o Framed panels shall be glazed with Grade 'A' safety glazing materials in accordance with table 3.13 except that normal glass complying with table 3.14 and having an area not greater than 0.5 m^2 may be used.

  o Unframed panels shall be glazed with Grade 'A' safety glazing material complying with both Table

3.13 and Table 3.17 with regard to the size requirements for the selected Grade 'A' safety glazing material.

- All unframed glazing in residential buildings where the lowest sight line of the glazing panel is less than 500 mm from the highest abutting finished floors level shall be in accordance with the table 3.13 and 3.17.
- Glass in shower doors, shower screens, bath enclosures, and associated windows the lowest sight line of which is less than 1500 mm above the highest abutting finished level of the floor, bottom of the bath, or shower base, shall be glazed with Grade 'A' safety glazing material in accordance with table 3.13.
- Framed panels that are not doors or side panels in internal partitions and shop fronts shall be glazed with Grade 'A' safety glazing material in accordance with table 3.13 except that those panels that can not be mistaken for a doorway or unimpeded path of travel may be glazed with normal glass in accordance with table 3.16 provided that the top edge is 1.5 m or greater above the highest abutting finished floor level and it can not be mistaken for a doorway or unimpeded path of travel.
- Panels which have the top and bottom edges framed and have one or more side edges unframed shall be glazed in accordance with table 3.17.
- Unframed glass panels which are not covered above shall be toughened safety glass in accordance with table 3.13.
- Grade 'A' safety glazing material in accordance with table 3.13, but not less than 6 mm standard nominal thickness, shall be used for all balustrade infill panels which are not required to support the handrail loads.
- Where glass used in school and child-care buildings which are primarily occupied by children under 16 years.

Table 3.13 Maximum Areas of Clear or Patterned Grade 'A' Safety Glazing Material for Framed Glass Doors, Framed Glass Side Panels, and Other Glazed Panels

Type of glass	Standard nominal thickness (mm)	Maximum area (m^2)
Tempered safety glass	3	1.0
	4	2.0
	5	3.0
	6	4.0
	8	6.0
	10	8.0
	12	10.0
Laminated safety glass	5.38	2.0
	6.38	3.0
	8.38	5.0
	10.38	7.0
	12.38	9.0
Where sealed insulating glass units are used the maximum areas specified in may be multiplied by 1.5 provided that each of the component glasses of the unit otherwise comply with the relevant requirements.		

Table 3.14 Maximum Areas of Clear or Patterned Normal Glass for Framed Glass Doors, Framed Glass Side Panels, and Other Framed Glass Side Panels, and Other Framed Glazed Panels

Standard nominal thickness (mm)	Maximum area (m^2)
3	0.1
4	0.3
5	0.5
6	0.9
8	1.8
10	2.7
12	4.5
15	6.3
19	8.5
25	12.0

Table 3.15 Maximum Areas of Grade 'A' Safety Glazing Materials for Shower Doors, Shower Screens and Bath Enclosures

Type of glass	Standard nominal thickness (mm)	Maximum area (m^2)
Safety wired glass	≥ 6	2.5

Table 3. 16 : Maximum Areas of Normal (Annealed) Glass for Shop Fronts, Internal Partitions, and Other Glazed Panels

Standard nominal thickness (mm)	Maximum area (m^2)		
	Framed		Top edge unframed
	Side panels and internal partitions	Shopfront side panels and Shopfronts	
3	0.8	0.8	0.1
4	1.4	1.4	0.3
5	2.2	2.2	0.5
6	3.3	3.3	0.9
8	4.5	6.0	1.8
10	6.0	9.0	2.7
12	8.0	12.0	4.5
15	10.0	15.0	6.3
19	12.0	15.0	8.5
25	15.0	15.0	12.0

Table 3.17: Shop Fronts, Internal Partitions, and Other Glazed Panels with Unframed Side Edges

Height of glass (m)	Type of glass	Minimum standard nominal thickness (mm)	Maximum number of vertical butt joints	Maximum panel width (m)
≤2	Normal	6.0	1	0.8
	Tempered	6.0	1	1.0
	Tempered	6.0	2	1.2
	Tempered	8.0	No limit	No limit
	Laminated	6.38	2	1.2
	Laminated	8.38	No limit	No limit
>2 ≤ 2.5	Normal	10.0	1	1.0
	Tempered	10.0	1	1.2
	Tempered	8.0	2	1.2
	Tempered	10.0	No limit	No limit
	Laminated	8.38	2	1.2
	Laminated	10.38	No limit	No limit
> 2.5 ≤2.8	Tempered	10.0	2	1.2
	Tempered	12.0	No limit	No limit
	Laminated	13.38	2	1.2
	Laminated	12.38	No limit	No limit
>2.8 ≤3.2	Tempered	12.0	2	1.2
	Laminated	12.38	2	1.2

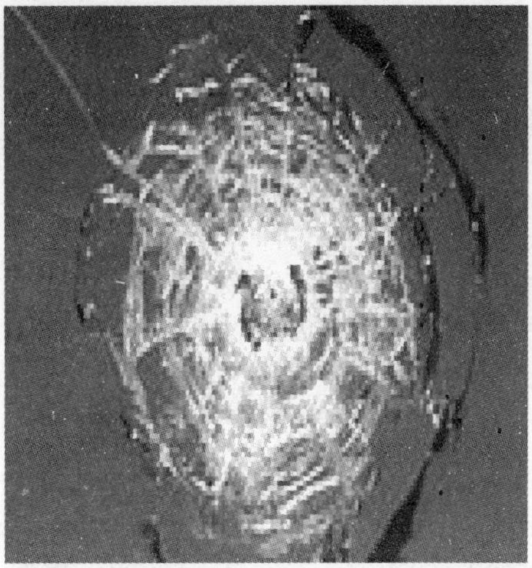

3.4.4 Security

Glass is generally regarded as a fragile and easily breakable material. To facilitate use of glass in buildings, security of life and property is an important parameter. Security glazing is used in situations where a high degree of protection either to persons or property is required, without any compromise on its optical properties. The main considerations for security are as follows:

3.4.4.1 Security from Burglary

Burglaries in residential and commercial buildings are common occurrences. Burglaries frequently are not directed at a specific facility but towards targets of easy opportunity and low perceived risk. The most critical step in a burglary is entry. One of the most common means of entry involves a window, either opened or broken to serve as a portal or to provide a means of unlocking an entrance door.

Standard two-ply laminated architectural glass provides a significant improvement over monolithic glass in resistance to forced entry from a variety of hand-held weapons - hammers, crowbars and bricks etc. Laminated glass, unlike normal glass panes cannot be cut from one side only, making quiet glasscutters useless as a burglary tool.

3.4.4.2 Security from Bullets

Security glazing can successfully resist both penetration of bullets and impact - induced spell fragmentation from the protected side of the glazing as a resulting of ballistic attacks.

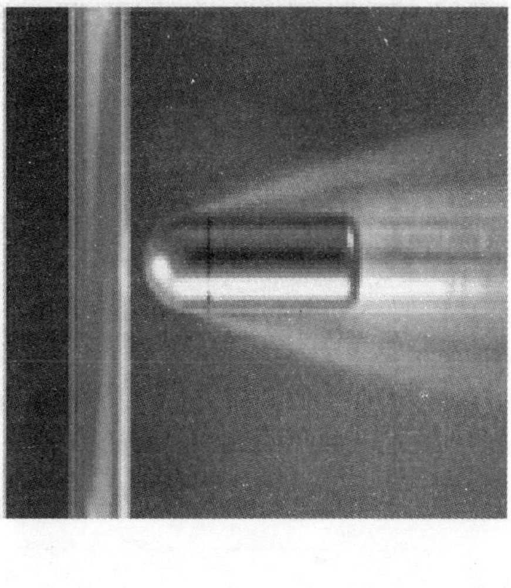

Laminated glass demonstrates the following features when subjected to blast loads:

- Laminated security glazing can continue to remain entirely in the frame at blast loads significantly greater than the load required to crack the lite.
- Laminated glass will crack with sufficient loading yet glass fragments tend to be retained by the interlayer unlike monolithic glass, which once cracked, will shatter and evacuate the frame.
- Laminated security glazing even as thin as 1/4 inch (6mm) will substantially extend the range of blast loads for which the glass is retained in the frame.

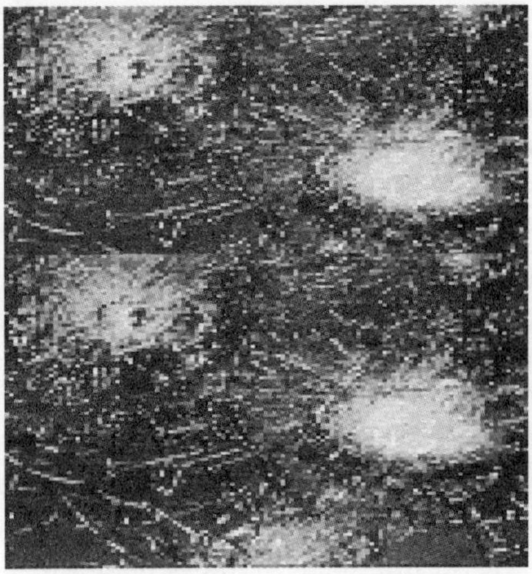

- Laminated security glazing will not generally shatter from impacts of small fragments propelled by the blast (i.e. nails, stones, etc.) even if the fragments penetrate the glazing.

3.4.4.3 Security from Explosions

In an explosion, glass causes most of the damage and injuries. If the glass e.q. laminated glass is confined in frame, there will be neither flying glass debris nor falling pieces of glass. Moreover, given that the facade is preserved, the interior damage will be substantially limited.

3.4.4.4 Security from Earthquakes

Laminated glass composed of one sheet of ordinary glass and one sheet of heat-tempered glass opens the way to an anti-earthquake glazing, particularly when used in combination with seismically engineered systems. The glass even if broken should stay in position.

3.4.4.5 Security from Hurricanes

During a storm, the most vulnerable parts of homes and buildings are the doors and windows. The glazings fitted with laminated glass resist the pressure of high winds for several hours, the impact of windborne debris of large dimensions, sudden gusts, changes in pressure and wind direction.

In zones exposed to violent storms, windows equipped with laminated glass are always recommended. Even if they break the glass shards will adhere to the interlayer, the glazing stays in place and the occupants remain protected.

Laminated glass can be designed to provide any specified degree of security. The precise constituents of the laminated glass will depend on the particular purpose and the specification should developed after detailed discussion with supplier, who should also be able to advise on the correct manner of glazing.

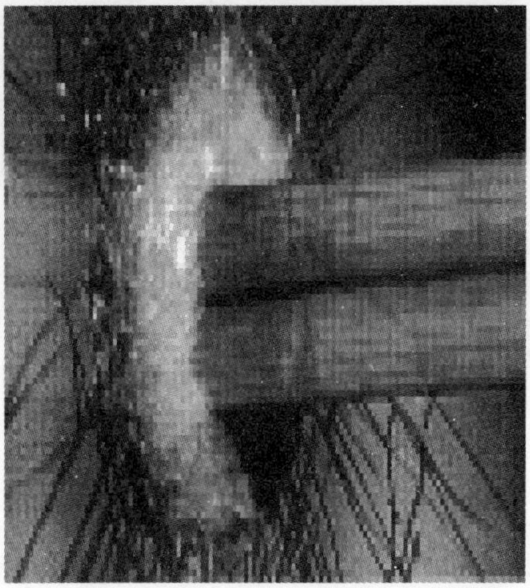

3.5 STRENGTH

India is divided into six zones with respect to wind velocities. Each zone has different wind velocities at different heights above the ground. For further details refer chapter 4. For wind load resistance, glass needs to be designed according to the situation, taking into account the load, the support conditions and the type of glass. The calculations are relatively simple and tables can be used to workout appropriate thickness of glass for windows / glazing.

The performance of associated fittings & fixtures such as handles, bolts etc. should also be considered along with the strength of the glass with a view that if the human body strikes against the glass, while operating the fitting & fixtures, the safety of the human body is ensured.

The strongest type of glass is tempered glass (4 times stronger) and is the only choice for doors and glass assemblies on patch fitting and bolts. Heat strengthened glass is next best in strength (2 times stronger); Laminated glass strength is same as that of a single glass of total thickness.

Normal glass is extremely dangerous in earthquakes and other natural disasters. Only safety glass (tempered or laminated) should be used. Certain buildings like hospitals, fire services, police stations etc. have much greater need to remain serviceable in such situations and should use laminated-tempered glass only.

3.6 AESTHETICS

Aesthetics play an important decisive role in deciding a particular glass. Aesthetic appearance of a building depends on:

- Colour of light reflected
- Amount of light reflected

Colour of light reflected depends on:

- Surrounding conditions (sky/ buildings/ vegetation etc.)
- Position, type and colour of coating
- Colour of base glass

Amount of light reflected depends on:

- Amount of incident light
- Type and position of coating

While from the technical standpoint, performance requirements and costs should narrow down the available choices, the selection is often influenced by the visual appeal of glass in interiors or exteriors. The selection of glass for a particular application from aesthetic point of view is largely a matter of designers preferences & owners choice.

3.7 GENERAL INFORMATION

The general information contains dimensional and framing requirements, Standard Nominal Thickness of Glass, Design Requirements for Specific Situations and Chair Rail.

3.7.1 Dimensional Requirements

Although some glass (particularly toughened glass) may be adequately designed from a structural strength point of view by the methods described in chapter 4.0, glasses of high strength may be flexible that they are unsatisfactory from an appearance point of view toughening does not increase stiffness. Edge movements associated with flexible glasses can result in excessive strains on the sealant, leading to seal failure and water leakage. For flexible high strength glasses, deflection limitations must always be considered as part of the overall design. It is recommended that the maximum deflection for all glass should be limited to span/60.

3.7.2 Framing Requirements

Frames when completely assembled and glazed, the frame member supporting the edge of the glass should not deflect more than the following values at the appropriate permissible stress design load:

(a) Span/150 for windows and sliding doors for residential applications irrespective of height and for other buildings less than 10 m high.

(b) Span/240 for other than residential applications greater than 10 m high.

A panel glazed directly into the building structure by means of appropriate beads or stops shall be considered to be framed provided that the assembly complies with the deflection requirements as mentioned above. Mixed framing - glass supported along the top and bottom edges by one means and along the vertical edges by another means shall be considered to be framed provided that the assembly complies with the deflection requirements.

3.7.3 Standard Nominal Thickness of Glass

Glass specified in these guidelines refers to various glass types having a standard nominal thickness.

Linear interpolation in determination of glass area or span with respect to glass thickness shall be carried out using the minimum thickness of the non-standard glass relative to the minimum thickness of the standard nominal thickness of the appropriate glass type.

3.7.4 Use of Glass in Specific Situations

The glass may be used as mirror, doors, side panels, shower doors, shower screens, bath enclosures, shop fronts, and internal partitions etc. It may strictly be required to use the toughened and laminated safety glasses in most of situations except in case of mirror. The relevant guidelines to use the glass in such situations are explained in following articles.

3.7.4.1 Use as Mirror

In order to use the glass as mirror, the recommendations with respect to thickness of glass & maximum permissible sizes given in the table 3.18 may be followed:

Table 3.18

Nominal Thickness (mm)	Maximum Size (mm)
3	1318 x 2438
5	1829 x 3048
6	2140 x 3050

3.7.4.2 Use as Doors

Glass used as doors shall be toughened and laminated safety glasses in accordance with table 3.13 with following exceptions:

- In fully framed panels, normal glass upto a maximum area of 0.5 m^2 may be used.
- Unframed doors shall be glazed with toughened safety glass with a SNT of not less than 10 mm

3.7.4.3 Use as Side Panels

The side panels may be used as framed as well as unframed. The requirements under both of these support conditions are described in following sub articles.

3.7.4.3.1 Framed Side Panels

The framed glass used in side panels with their nearest vertical sight line less than 300 mm from the nearest edge of the doorway opening shall be a toughened and laminated safety glasses in accordance with table 3.13.

3.7.4.3.2 Unframed Side Panels

Unframed glass in side panels without exposed edges shall be glazed with toughened or laminated safety glasses in accordance with the requirements shown in table 3.17.

3.7.5 Use as Shower Doors, Shower Screens and Bath Enclosures

All glass in shower doors, shower screens, bath enclosures, and associated windows the lowest sight lines of which is less than 1500 mm above the highest abutting finished level of the floor, bottom of the bath, or shower base, shall be glazed with toughened and laminated safety glasses in accordance with table 3.13.

3.7.6 Use as Shop-fronts and Internal Partitions

The shop front and internal partition can be used framed as well as unframed as described below:

3.7.6.1 Framed Panels

Framed panels that are not doors or side panels in internal partition and shop-fronts shall be glazed with toughened and laminated safety glasses in accordance with table 3.13 except that those panels that cannot be mistaken for a doorway or unimpeded path of travel may be glazed with annealed glass in accordance with table 3.16.

3.7.6.2 Unframed Panels

The panels may be unframed either at top or side edges. The requirements under these situations are as follows:

3.7.6.2.1 Top Edge Unframed

Panels that are framed on three sides but not on the top edge shall be glazed with toughened and laminated safety glasses in accordance with table 3.13 or with normal glass in accordance with table 3.16, provided that the top edge is 1.5m or greater above the highest abutting finished floor level and it cannot be mistaken for a doorway or unimpeded path of travel.

3.7.6.2.2 Side Edges Unframed

Panels having the top and bottom edges as framed and one or more side edges unframed shall be glazed in accordance with table 3.17.

3.7.6.2.3 Other Unframed Panels

Unframed glass panels, which are not covered by articles 3.7.6.2.1 and 3.7.6.2.2, shall be of toughened safety glass in accordance with table 3.13.

3.7.7 Shelves

The maximum evenly distributed safe load that a shelf can support is dependent on glass type, thickness, width and the span of the glass between supports. The maximum evenly distributed safe load for a rectangular shelf can be calculated from the following equation:

Maximum evenly distributed safe load (kg) = XLD

Where, X = maximum evenly distributed safe load per unit area in kg/m^2 from table 3.19.

L = the unsupported span of the shelf in metres.

D = the depth of the shelf in metres.

- Normal glass is acceptable for use as shelves only when fully enclosed within a cabinet.
- Glass shelves that are not fully enclosed in a cabinet shall be laminated glass / tempered glass.
- The glass shelves should overlap the support by at least 4 mm.

Table 3.19 Maximum evenly distributed safe load per unit area supported by a glass shelf for determining safe loading capacity (kg/m²) for a given unsupported length, thickness and type of glass.

Type of glass	Nominal thickness t of glass (mm)	X, the maximum evenly distributed safe load per unit area (kg/m²) supported by the following lengths L of glass in mm												
		300	400	500	600	700	800	900	1000	1100	1200	1300	1400	1500
Normal	4	153	86	55	38	28	21	17	11	14	10	8	7	6
	5	244	137	88	61	45	34	27	22	18	15	13	11	10
	6	356	200	128	89	65	50	40	32	26	22	19	16	14
	8	627	353	226	157	41	88	70	56	47	39	33	29	25
	10	995	559	358	249	183	140	111	90	74	62	53	46	40
	12	1447	814	521	362	266	203	161	130	108	90	77	66	58
Laminated	4	69	39	25	17	13	10	8	6	5	4	4	3	3
	6	166	93	60	41	30	23	18	15	12	10	9	8	7
	8	305	172	110	76	56	43	34	27	23	19	16	14	12
	10	487	274	175	122	89	69	54	44	36	30	26	22	19
	12	711	400	256	178	131	100	79	64	53	44	38	33	28
Tempered	4	988	417	213	123	78	52	37	27	20	15	12	10	8
	5	1991	840	430	249	157	105	74	54	40	31	24	20	16
	6	2997	1482	759	439	276	185	130	95	71	55	43	35	28
	8	5283	2972	1775	1027	647	433	304	222	167	128	101	81	66
	10	8383	4716	3018	2054	1293	866	608	444	333	257	202	162	131
	12	12197	6861	4391	3049	2240	1520	1068	778	585	450	354	284	231

Guarding loads: as per BS 6180: 1995

	Line load UDL KN/m run	UDL (infill) KN/ sq m	Point Load KN
Office	0.74	1	0.5
Industrial & Stores light traffic	0.36	0.5	0.25
Industrial & stores heavy traffic	0.74	1	0.5
Library, museum, non residential	0.74	1	0.5
Hospital, Schools & Colleges	0.74	1	0.5
Sports ground	3	1.5	1.5
Retail	1.5	1.5	1.5
Cinema, disco & points of assembly	3.0	1.5	1.5
Residential	0.36	0.5	0.25
Hotels & Motels	0.74	1	0.5
Hostels & guest houses	0.74	1	0.5

3.7.8 Glass as Guarding

When glass is used to protect a change in level whether as part of balustrade, screen or window, it is said to be acting as guarding. For domestic applications guarding is to be provided to protect changes in level greater than 600 mm. In commercial applications, guarding is to be provided were there is a change in level of 380mm or the equivalent of two stair risers.

In all buildings, where glazing is protecting a change of level and the glass is within 800mm of finish floor level, it must have sufficient strength to "provide containment". Where barriers are to be used in dwelling and other buildings in which children will be present, consideration of the design should be made to ensure that any gaps are less than 100mm and that the barrier cannot be easily climbed. Line load (KN/m run), infill load (KN/sq m) and point load (KN) are used in the design of barriers incorporating glass depending on their location within a building and the building usage.

The loads are to be considered separately and are not cumulative. The design should take into account the most unfavorable likely imposed loads without excessive deflections or distortion.

3.7.9 Chair Rail

A glazing bar or rigid push bar in a glazed window/ door, which provides protection from human impact, is known as chair rail.

Chair rails shall have a width in the vertical direction of not less than 40 mm and shall be designed for the horizontal railing loads. Where the chair rail is installed clear of the glazing, it shall also be designed for the vertical railing loads and the deflection of the chair rail under the horizontal loads shall be such that the chair rail remains clear of the glass panel under 1.5 times the specified rail loads.

3.7.10 Design Requirements for Other Situations

Where glass is used in situations not otherwise covered in these guidelines, the manufacturers shall be consulted for design data. But in all cases, consideration shall be given to the consequences of glass breakage, and the design criteria shall be appropriately chosen to suit the application.

Detailed properties of normal glass, laminated glass, tempered glass, reflective glass, insulating glass and mirror along with their commonly used variants are described in this chapter. Selection of a specific glass type depends upon a number of building requirements such as solar control, sound insulation, thermal strength, fire resistance, safety, security strength & aesthetic appeal. The chapter covers each of these requirements to assist the reader for specific selection of glass type.

4

Determination of Appropriate Glass Thickness

4.0 DETERMINATION OF APPROPRIATE THICKNESS OF GLASS

4.1 GENERAL

Once the type of glass to be used is selected, this chapter deals with the determination of appropriate thickness of glass. The thickness of the glass to be used in window panels is governed by the following factors:

i. Area to be covered by the window panel.

ii. Support conditions (supported on two sides or four sides).

iii. Aspect ratio of window panel (length / breadth).

iv. Effective wind pressure at the window height

v. Strength/load bearing capacity of glass to be used.

A simplified procedure is described in the following sections for the determination of appropriate thickness of glass in a window panel. In general, the provisions of Australian Code AS 1288-1994, British Code BS 6262 (Part-3) - 1996 and IS 3548-1988 have been considered. The wind pressure computation is based on the procedure given in IS: 875 (Part –3) 1987, (Reaffirmed 1997).

The salient features of the procedure are as follows:

- It is for windows in walls of rectangular buildings with height to width ratio less than or equal to six ($H/B \leq 6$).
- The maximum area of glass panel is restricted to 15 m^2.
- The maximum span of window is restricted to 4 m.
- Aspect ratio of the glass panel should be greater than 1.5. If it is less than 1.5, next higher available thickness should be selected.
- The factor of safety used is 2.5 considering the variability in strength of glass.
- Applicable to normal, reflective, laminated, tempered and insulating glass.

- Applicable to rectangular panels properly fixed/ grouted.
- Design minimum thickness of the glass will be the maximum value of the thickness arrived at under article 4.3, 4.4 and 4.5.

For any other condition with respect to wind load not covered above, appropriate clauses of IS: 875 (Part – 3) –1987 (Reaffirmed 1997) should be referred.

The numerical examples for computation of glass thickness for various locations and glass specifications are given in the article 4.6.

4.2 STANDARD NOMINAL THICKNESS (SNT)

The glass sheets may have different thickness at different locations. Table 4.1 shows acceptable thickness limits for different types of the glass.

Table 4.1 Thickness Limits for Standard Nominal Thickness of Glass (mm)

Type of glass	Nominal thickness	Thickness limits	
		Minimum	Maximum
Normal Reflective& Tempered glass	3	2.8	3.2
	4	3.8	4.2
	5	4.8	5.2
	6	5.8	6.2
	8	7.7	8.3
	10	9.7	10.3
	12	11.7	12.3
	15	14.5	15.5
	19	18.0	20.0
	25	23.5	26.5
Laminated glass	5.38	4.95	5.81
	6.38	5.95	6.81
	8.38	7.95	8.81
	10.38	9.95	10.81
	12.38	11.95	12.81
	16.38	15.75	17.01

4.3 DESIGN CONSIDERATIONS

This article deals with the design strength of the glass and the empirical relation to evaluate the minimum thickness or the maximum allowable area of the glass panel for a particular glass type for given set of design wind pressure, support condition and aspect ratio.

4.3.1 Design Flexural Tensile Strength of Glass

The required thickness of the glass depends upon the design strength obtained after applying a factor of safety of 2.5. The minimum design strength of normal glass for thickness $\leq$ 6mm is 16.7 N/mm^2 and for thickness > 6mm is 15.2 N/mm^2.

4.3.2 Empirical Relationship

Assuming that the normal glass has design strength in accordance with article 4.3.1, following empirical relation between the wind pressure, area of the glass panel and the required glass thickness can be used:

$$P_{net} * A = 200.0 * T^k \quad (T \leq 6 \text{ mm}) \qquad \dots (4.1)$$

$$P_{net} * A = 200.0 * T^k + 1900 \quad (T > 6 \text{ mm}) \qquad \dots (4.2)$$

where,

P_{ne} = Net design wind pressure (N/m^2) as per article 4.5

A = area of glass panel (m^2)

T = SNT of the normal glass (mm),

k = a constant as shown in table 4.2.

Table 4.2 SNT and Corresponding 'k' Values

'T'	'k'	'T'	'k'
3 mm	1.683	10 mm	1.578
4 mm	1.732	12 mm	1.583
5 mm	1.753	15 mm	1.579
6 mm	1.765	19 mm	1.569
8 mm	1.570	25 mm	1.569

4.3.3 Limiting Aspect Ratio (AR_{max})

The design of the thickness using empirical relation in accordance with article 4.3.2 will be valid upto a limiting aspect ratio AR_{max}. The value of AR_{max} for different SNT of glass is shown in the following table 4.3.

Table 4.3 SNT and Corresponding AR_{max} Values

SNT	AR_{max}	SNT	AR_{max}
3 mm	7.3	10 mm	4.9
4 mm	6.8	12 mm	4.3
5 mm	6.5	15 mm	3.8
6 mm	6.3	19mm	3.3
8 mm	5.9	25 mm	2.9

4.4 DETERMINATION OF MINIMUM GLASS THICKNESS

The determination of minimum glass thickness of panels of different types of glass supported on four sides as well as two opposite sides is discussed in the following sections.

4.4.1 Glass Panels Supported on All Four Sides

The thickness of panels of normal glass can be obtained using the empirical relation explained in article 4.3.2. This relation is valid for the glass having the minimum design strength of the material as mentioned in article 4.3.1 and the maximum aspect ratio limited to a value in accordance with article 4.3.3. However, if the aspect ratio exceeds the prescribed values under article 4.3.3, the design will be carried out in accordance with the article 4.4.2 applicable to glass supported on two opposite sides.

4.4.1.1 Normal (Annealed)/Reflective Glass

The minimum thickness of normal/reflective glass for a particular value of net design wind pressure P_{net} can be evaluated as per the procedure given in article 4.3. User can refer table 4.13 or the figure 4.1 to directly obtain the minimum thickness or the allowable maximum area of the glass panel.

4.4.1.2 Laminated/Tempered/Insulating Glass

To determine the thickness of laminated / tempered / insulating glass, the design wind pressure P_{net} is modified as below:

$$P_{net} = P_{net}/P_f$$

where, P_f is the pressure factor dependent on the type of glass.

The values of the pressure factor P_f can be taken from table 4.4 as below:

Table 4.4 Glass Types and Pressure Factors

S. No	Glass type	P_f
1.	Normal (Annealed)	1.00
2.	Laminated	0.80
3.	Tempered	2.5
4.	Insulating 1.50	

Using the modified value of P_{net} as explained in this article, the thickness of other types of glass can be obtained in accordance with article 4.3. However, tables 4.14, 4.15, 4.16 or figures 4.2, 4.3, 4.4 may be referred to directly get the minimum thickness or the allowable maximum area of glass panel for laminated, tempered and insulated glass respectively.

4.4.2 Glass Supported on Two Opposite Sides

Normal and laminated glass panels supported on two opposite sides can be designed using following empirical relations:

For $T \leq 6$ mm

$$b = \frac{4.39 \times T}{\sqrt{P_{net}/P_f}} \qquad \ldots(4.3)$$

For $T > 6$ mm

$$b = \frac{4.22 \times T}{\sqrt{P_{net}/P_f}} \quad \ldots(4.4)$$

Tempered glass panels supported on two opposite sides can be designed using following empirical relations:

For T 6 mm

$$b = \frac{3.3688 \times T}{\sqrt{P_{net}/P_f}} \quad \ldots(4.5)$$

For $T > 6$ mm

$$b = \frac{2.9069 \times T}{\sqrt{P_{net}/P_f}} \quad \ldots(4.6)$$

Where,

b = Span in meters

P_{net} = Net design wind pressure in N/m^2 as per article 4.5

P_f = Pressure factor as given in Table 4.4.

T = SNT of glass (mm)

- For laminated glass, thickness of PVB, 0.38 mm should be reduced i.e. T = SNT – 0.38
- For insulating glass thickness, of only one glass pane should be considered. If the glass panes are of different thickness, the minimum of the two thicknesses should be considered.

Above empirical relation may be used to calculate thickness of different types of glass for given pressure and span.

Table 4.17 / figure 4.5 for normal glass, table 4.18 / figure 4.6 for laminated glass and table 4.19 / figure 4.7 for tempered glass may also be used directly.

4.5 NET DESIGN WIND PRESSURE (P_{NET})

Net design wind pressure (P_{net}) is an important parameter governing the thickness of glass to be used in the window panels. It depends on several factors i.e. location of building (wind zone), construction patterns around buildings (terrain category), topography of site, building plan and height etc.

Net design wind pressure (P_{net}), may be defined using the following equation:

$$P_{net} = P_z \times C_p \qquad \ldots(4.7)$$

where, C_p = Net pressure coefficient

P_z = Design wind pressure at height 'z' (N/m^2)

Important terms related to the equation 4.7 are defined in the following sections.

4.5.1 Wind Zone

India has been divided into six wind zones. Each wind zone has a specific basic wind speed. These basic wind speeds have been used to compute design wind pressures. A proper wind zone should be selected for the site of a building. For ready reference, wind zones and basic wind speeds of some important cities and towns are given in the table 4.5.

4.5.2 Terrain Category

Wind pressure also depends on terrain type where building is to be located. Terrain category means the characteristics of the surface irregularities of an area, which arise from natural or constructed features. Wind pressures have been computed for all the four terrain categories which are as follows:

Category I – Exposed open terrain with few or no obstructions and in which the average height of any object surrounding the structure is less than 1.5m.

Table 4.5: Basic Wind Speep at 10m Height for Some Important Cities/Towns

City / Town	Wind Zone	Basic Wind Speed (m/s)	City / Town	Wind Zone	Basic Wind Speed (m/s)
Agra	IV	47	Jhansi	IV	47
Ahmadabad	II	39	Jodhpur	IV	47
Ajmer	IV	47	Kanpur	IV	47
Almora	IV	47	Kohima	III	44
Amritsar	IV	47	Kurnool	II	39
Asansol	IV	47	Lakshadweep	II	39
Aurangabad	II	39	Lucknow	IV	47
Bahraich	IV	47	Ludhiana	IV	47
Bangalore	I	33	Madras	V	50
Barauni	IV	47	Madurai	II	39
Bareilly	IV	47	Mandi	II	39
Bhatinda	IV	47	Mangalore	II	39
Bhilai	II	39	Moradabad	IV	47
Bhopal	II	39	Mysore	I	33
Bhubaneshwar	V	50	Nagpur	III	44
Bhuj	V	50	Nainital	IV	47
Bikaner	IV	47	Nasik	II	39
Bokaro	IV	47	Nellore	V	50
Bombay	III	44	Panjim	II	39
Calcutta	V	50	Patiala	IV	47
Calicut	II	39	Patna	IV	47
Chandigarh	IV	47	Pondicherry	V	50
Coimbatore	II	39	Port Blair	III	44
Cuttack	V	50	Pune	II	39
Darbhanga	VI	55	Raipur	II	39
Darjeeling	IV	47	Rajkot	II	39
Dehra Dun	IV	47	Ranchi	II	39
Delhi	IV	47	Roorkee	II	39
Durgapur	IV	47	Rourkela	II	39
Gangtok	IV	47	Simla	II	39
Gauhati	V	50	Srinagar	II	39
Gaya	II	39	Surat	III	44
Gorakhpur	IV	47	Tiruchchirrappalli	IV	47
Hyderabad	III	44	Trivandrum	II	39
Imphal	IV	47	Udaipur	IV	47
Jabalpur	IV	47	Vadodara	III	44
Jaipur	IV	47	Varanasi	IV	47
Jamshedpur	IV	47	Vijaywada	V	50

Note: - For any city not covered in this table IS : 875 (Part 3) 1987 may be referred.

Category II - Open terrain with well-scattered obstructions having heights generally between 1.5-10m.

Category III – Terrain with numerous closely spaced obstructions having the size of building structures up to 10m height with or without a few isolated tall structures.

Category IV – Terrain with numerous large, high and closely spaced obstructions.

Design wind pressures have been computed and given in tables 4.6 to 4.11.

4.5.3 Class of Structure

The buildings/structures are classified into the following three different classes depending upon their size:

Class A – Structures and/or their components such as cladding, glazing, roofing etc., having maximum dimension (greatest horizontal or vertical dimension) less than 20m.

Table 4.6 Design Wind Pressure 'P_z' for Wind Zone-1 **(Basic wind speed 33 m/sec)**

Height (m)	Design Wind Pressure (N/m²)			
	Terrain Category I	Terrain Category II	Terrain Category III	Terrain Category IV
5.0	794.21	720.37	596.54	461.04
10.0	794.21	720.37	596.54	461.04
15.0	855.88	794.21	677.80	461.04
20.0	903.64	824.76	734.85	461.04
25.0	928.00	863.75	771.68	564.21
30.0	952.69	903.64	809.41	677.80
35.0	973.52	923.92	832.48	723.98
40.0	994.57	944.43	855.88	771.68
45.0	1015.84	965.16	879.59	820.91
50.0	1037.34	986.12	903.64	871.65
60.0	1058.19	1009.86	929.64	903.64
70.0	1079.25	1033.88	956.01	936.20
80.0	1100.51	1058.19	982.75	969.33
90.0	1121.99	1082.78	1009.86	1003.05
100.0	1143.67	1107.65	1037.34	1037.34
125.0	1180.26	1143.67	1072.20	1072.20
150.0	1217.43	1180.26	1107.65	1107.65
175.0	1236.23	1198.77	1134.61	1134.61
200.0	1255.18	1217.43	1161.89	1161.89
250.0	1293.50	1255.18	1198.77	1180.26

Table 4.7 Design Wind Pressure 'P_z' for Wind Zone- II (Basic wind speed 39 m/sec)

Height (m)	Design Wind Pressure (N/m²)			
	Terrain Category I	Terrain Category II	Terrain Category III	Terrain Category IV
5.0	1130.50	1025.40	849.13	656.25
10.0	1130.50	1025.40	849.13	656.25
15.0	1218.28	1130.50	964.80	656.25
20.0	1286.26	1173.98	1046.01	656.25
25.0	1320.94	1229.48	1098.43	803.12
30.0	1356.09	1286.26	1152.14	964.80
35.0	1385.73	1315.13	1184.98	1030.53
40.0	1415.69	1344.32	1218.27	1098.43
45.0	1445.97	1373.83	1252.04	1168.50
50.0	1476.57	1403.67	1286.26	1240.73
60.0	1506.25	1437.46	1323.27	1286.26
70.0	1536.23	1471.65	1360.81	1332.61
80.0	1566.50	1506.25	1398.87	1379.78
90.0	1597.06	1541.25	1437.46	1427.76
100.0	1627.92	1576.65	1476.57	1476.57
125.0	1680.01	1627.92	1526.20	1526.20
150.0	1732.92	1680.01	1576.65	1576.65
175.0	1759.68	1706.36	1615.03	1615.03
200.0	1786.65	1732.92	1653.86	1653.86
250.0	1841.20	1786.65	1706.36	1680.01

Table 4.8 Design Wind Pressure 'P_z' for Wind Zone- III (Basic wind speed 44 m/sec)

Height (m)	Design Wind Pressure (N/m²)			
	Terrain Category I	Terrain Category II	Terrain Category III	Terrain Category IV
5.0	1466.23	1329.92	1101.30	851.15
10.0	1466.23	1329.92	1101.30	851.15
15.0	1580.07	1466.23	1251.32	851.15
20.0	1000.25	1522.02	1356.65	851.15
25.0	1713.23	1594.60	1424.64	1041.62
30.0	1758.81	1668.25	1494.29	1251.32
35.0	1797.26	1705.69	1536.88	1336.57
40.0	1836.12	1743.55	1580.07	1424.64
45.0	1875.39	1781.83	1623.86	1515.51
50.0	1915.08	1820.52	1668.25	1609.20
60.0	1953.57	1864.35	1716.25	1668.25
70.0	1992.45	1908.70	1764.94	1728.36
80.0	2031.71	1953.57	1814.30	1789.54
90.0	2071.35	1998.97	1864.35	1851.78
100.0	2111.38	2044.88	1915.08	1915.08
125.0	2178.93	2111.38	1979.45	1979.45
150.0	2247.56	2178.93	2044.88	2044.88
175.0	2282.27	2213.11	2094.65	2094.65
200.0	2317.25	2247.56	2145.02	2145.02
250.0	2388.00	2317.25	2213.11	2178.93

Table 4.9 Design Wind Pressure 'P_z" for Wind Zone -IV (Basic wind speed 47m/sec)

Height (m)	Design Wind Pressure (N/m²)			
	Terrain Category I	Terrain Category II	Terrain Category III	Terrain Category IV
5.0	1672.99	1517.45	1256.60	971.17
10.0	1672.99	1517.45	1256.60	971.17
15.0	1802.88	1672.99	1427.77	971.17
20.0	1903.49	1737.33	1547.95	971.17
25.0	1954.82	1819.46	1625.53	1188.51
30.0	2006.83	1903.49	1705.01	1427.77
35.0	2050.69	1946.22	1753.60	1525.05
40.0	2095.03	1989.42	1802.88	1625.53
45.0	2139.84	2033.09	1852.85	1729.22
50.0	2185.13	2077.24	1903.49	1836.12
60.0	2229.05	2127.25	1958.26	1903.49
70.0	2273.41	2177.85	2013.82	1972.08
80.0	2318.20	2229.05	2070.14	2041.88
90.0	2363.44	2280.84	2127.25	2112.90
100.0	2409.11	2333.23	2185.13	2185.13
125.0	2486.19	2409.11	2258.57	2258.57
150.0	2564.49	2486.19	2333.23	2333.23
175.0	2604.10	2525.19	2390.02	2390.02
200.0	2644.01	2564.49	2447.50	2447.50
250.0	2724.73	2644.00	2525.19	2486.19

Table 4.10 Design Wind Pressure 'P_z' for Wind Zone- V (Basic wind speed 50 m/sec)

Height (m)	Design Wind Pressure (N/m²)			
	Terrain Category I	Terrain Category II	Terrain Category III	Terrain Category IV
5.0	1928.93	1749.60	1448.84	1119.74
10.0	1928.93	1749.60	1448.84	1119.74
15.0	2078.70	1928.93	1646.20	1119.74
20.0	2194.70	2003.12	1784.77	1119.74
25.0	2253.88	2097.81	1874.22	1370.33
30.0	2313.85	2194.70	1965.85	1646.20
35.0	2364.42	2243.96	2021.88	1758.36
40.0	2415.54	2293.77	2078.70	1874.22
45.0	2467.21	2344.13	2136.31	1993.77
50.0	2519.42	2395.03	2194.70	2117.02
60.0	2570.07	2452.69	2257.85	2194.70
70.0	2621.21	2511.03	2321.90	2273.78
80.0	2672.86	2570.07	2386.85	2354.26
90.0	2725.01	2629.78	2452.69	2436.14
100.0	2777.67	2690.19	2519.42	2519.42
125.0	2866.55	2777.67	2604.11	2604.11
150.0	2956.82	2866.55	2690.19	2690.19
175.0	3002.49	2911.51	2755.66	2755.66
200.0	3048.50	2956.82	2821.93	2821.93
250.0	3141.58	3048.50	2911.51	2866.55

Design:

All the steps explained under example 1 will be followed and some of the important calculations are being shown below:

Design wind pressure from table 4.9 for terrain category II is 2033 N/m^2

Coefficient C_p = 1.7

Net Pressure

$P_{net} = P_z \times C_p$

$= 2033 \times 1.7 = 3456$ N /m^2

From table 4.15, it is clear that the thickness of 8 mm tempered glass is sufficient to cater a pressure of 3456 N/ m^2 for 4.5 m^2 panel area.

Example 4 :

Same as in Example 1 with a change in the location of the glass panel at a height of 20 m.

Design:

All the steps explained in example 1 will be followed and some of the important calculations are being shown below:

From table 4.9 the value of P_z = 1737 N/ m^2.

From table 4.12, Coefficient C_p = 1.7

Therefore, $P_{net} = P_z \times C_p$

$= 1737 \times 1.7$

$= 2952.9$ N /m^2

From table 4.15, the maximum allowable area for 8 mm thick tempered glass panel is 5.90m^2, which is greater than the area of 4.5m^2 glass panel. Hence, 8 mm thick tempered glass panel can be used at this height also.

Example 5:

Same as Example 1 with a change in the location of the glass panel at a height 10 m.

Design:

All the steps explained under example 1 will be followed and some of the important calculations are being shown below:

From the table 4.9, P_z = 1517 N/ m^2

From the table 4.12, Coefficient C_p = 1.7

$P_{net} = P_z \times C_p \quad = 1517x\ 1.7 = 2578.9\ N/m^2$

From table 4.15, it is clear that the required thickness of the tempered glass for 4.5 m^2 panel will be 6 mm.

Considering above examples, a summary of the required thickness of tempered glass for glass panel sizes of 3.0 x 1.5m fixed on all sides is shown in the following table:

Height of window above the ground level	Glass types	SNT
Height 10 m	Tempered	6 mm
Height 20 m	Tempered	8 mm
Height 45 m	Tempered	8 mm
Height 60 m	Tempered	8 mm

In order to use normal glass instead of tempered glass, the thickness of glass panel comes out as given in the following table.

Height of window above the ground level	Glass types	SNT
Height 10 m	Normal	12 mm
Height 20 m	Normal	15 mm
Height 45 m	Normal	15 mm

From the above exercise it is inferred that the required thickness of the glass panel may reduce with the reduction in the height as well as reduction in size of the window panel.

This Chapter is a quick reference to assist the reader to decide the thickness of the glass type as a function of wind pressure on the glazing.

Table 4.13 : Maximum Areas (m^2) for Normal (Annealed) Glass Fixed on All Four Sides.

Design Wind Pr. (N/m^2)	Standard Nominal Thickness of Glass in mm									
	3.00	4.00	5.00	6.00	8.00	10.00	12.00	15.00	19.00	25.00
500	2.540	4.410	6.720	9.450	14.270	15.000	15.000	15.000	15.000	15.000
550	2.310	4.010	6.110	8.590	12.970	15.000	15.000	15.000	15.000	15.000
600	2.120	3.680	5.600	7.880	11.890	15.000	15.000	15.000	15.000	15.000
650	1.950	3.400	5.170	7.270	10.980	14.570	15.000	15.000	15.000	15.000
700	1.820	3.150	4.800	6.750	10.190	13.530	15.000	15.000	15.000	15.000
750	1.690	2.940	4.480	6.300	9.510	12.630	15.000	15.000	15.000	15.000
800	1.590	2.760	4.200	5.910	8.920	11.840	15.000	15.000	15.000	15.000
850	1.490	2.600	3.950	5.560	8.390	11.140	14.260	15.000	15.000	15.000
900	1.410	2.450	3.730	5.250	7.930	10.520	13.460	15.000	15.000	15.000
950	1.340	2.320	3.540	4.970	7.510	9.970	12.760	15.000	15.000	15.000
1000	1.270	2.210	3.360	4.730	7.130	9.470	12.120	15.000	15.000	15.000
1050	1.210	2.100	3.200	4.500	6.790	9.020	11.540	15.000	15.000	15.000
1100	1.160	2.010	3.050	4.300	6.490	8.610	11.020	14.810	15.000	15.000
1150	1.100	1.920	2.920	4.110	6.200	8.230	10.540	14.170	15.000	15.000
1200	1.060	1.840	2.800	3.940	5.950	7.890	10.100	13.580	15.000	15.000
1250	1.020	1.770	2.690	3.780	5.710	7.580	9.690	13.030	15.000	15.000
1300	0.980	1.700	2.580	3.640	5.490	7.280	9.320	12.530	15.000	15.000
1350	0.940	1.630	2.490	3.500	5.280	7.010	8.980	12.070	15.000	15.000
1400	0.910	1.580	2.400	3.380	5.100	6.760	8.660	11.640	15.000	15.000
1450	0.880	1.520	2.320	3.260	4.920	6.530	8.360	11.230	15.000	15.000
1500	0.850	1.470	2.240	3.150	4.760	6.310	8.080	10.860	14.800	15.000
1550	0.820	1.420	2.170	3.050	4.600	6.110	7.820	10.510	14.320	15.000
1600	0.790	1.380	2.100	2.950	4.460	5.920	7.570	10.180	13.870	15.000
1650	0.770	1.340	2.040	2.860	4.320	5.740	7.340	9.870	13.450	15.000
1700	0.750	1.300	1.980	2.780	4.200	5.570	7.130	9.580	13.060	15.000
1750	0.730	1.260	1.920	2.700	4.080	5.410	6.920	9.310	12.680	15.000
1800	0.710	1.230	1.870	2.630	3.960	5.260	6.730	9.050	12.330	15.000
Maximum Aspect Ratio	**7.3**	**6.8**	**6.5**	**6.3**	**5.9**	**4.9**	**4.3**	**3.8**	**3.3**	**2.9**

Contd...

	3.00	4.00	5.00	6.00	8.00	10.00	12.00	15.00	19.00	25.00
1850	0.690	1.190	1.820	2.550	3.860	5.120	6.550	8.810	12.000	15.000
1900	0.670	1.160	1.770	2.490	3.760	4.980	6.380	8.570	11.680	15.000
1950	0.650	1.130	1.720	2.420	3.660	4.860	6.210	8.350	11.380	15.000
2000	0.640	1.100	1.680	2.360	3.570	4.730	6.060	8.150	11.100	15.000
2050	0.620	1.080	1.640	2.310	3.480	4.620	5.910	7.950	10.830	15.000
2100	0.610	1.050	1.600	2.250	3.400	4.510	5.770	7.760	10.570	15.000
2150	0.590	1.030	1.560	2.200	3.320	4.400	5.640	7.580	10.320	15.000
2200	0.580	1.000	1.530	2.150	3.240	4.300	5.510	7.400	10.090	15.000
2250	0.560	0.980	1.490	2.100	3.170	4.210	5.390	7.240	9.860	14.720
2300	0.550	0.960	1.460	2.050	3.100	4.120	5.270	7.080	9.650	14.400
2350	0.540	0.940	1.430	2.010	3.040	4.030	5.160	6.930	9.440	14.090
2400	0.530	0.920	1.400	1.970	2.970	3.950	5.050	6.790	9.250	13.800
2450	0.520	0.900	1.370	1.930	2.910	3.860	4.950	6.650	9.060	13.520
2500	0.510	0.880	1.340	1.890	2.850	3.790	4.850	6.520	8.880	13.250
2550	0.500	0.870	1.320	1.850	2.800	3.710	4.750	6.390	8.700	12.990
2600	0.490	0.850	1.290	1.820	2.740	3.640	4.660	6.270	8.540	12.740
2650	0.480	0.830	1.270	1.780	2.690	3.570	4.570	6.150	8.380	12.500
2700	0.470	0.820	1.240	1.750	2.640	3.510	4.490	6.030	8.220	12.270
2750	0.460	0.800	1.220	1.720	2.590	3.440	4.410	5.920	8.070	12.040
2800	0.450	0.790	1.200	1.690	2.550	3.380	4.330	5.820	7.930	11.830
2850	0.450	0.770	1.180	1.660	2.500	3.320	4.250	5.720	7.790	11.620
2900	0.440	0.760	1.160	1.630	2.460	3.270	4.180	5.620	7.650	11.420
2950	0.430	0.750	1.140	1.600	2.420	3.210	4.110	5.520	7.520	11.230
3000	0.420	0.740	1.120	1.580	2.380	3.160	4.040	5.430	7.400	11.040
3050	0.420	0.720	1.100	1.550	2.340	3.100	3.970	5.340	7.280	10.860
3100	0.410	0.710	1.080	1.520	2.300	3.050	3.910	5.260	7.160	10.680
3150	0.400	0.700	1.070	1.500	2.260	3.010	3.850	5.170	7.050	10.510
3200	0.400	0.690	1.050	1.480	2.230	2.960	3.790	5.090	6.940	10.350
3250	0.390	0.680	1.030	1.450	2.200	2.910	3.730	5.010	6.830	10.190
3300	0.390	0.670	1.020	1.430	2.160	2.870	3.670	4.940	6.730	10.040
Maximum Aspect Ratio	**7.3**	**6.8**	**6.5**	**6.3**	**5.9**	**4.9**	**4.3**	**3.8**	**3.3**	**2.9**

Contd...

	3.00	4.00	5.00	6.00	8.00	10.00	12.00	15.00	19.00	25.00
3350	0.380	0.660	1.000	1.410	2.130	2.830	3.620	4.860	6.630	9.890
3400	0.370	0.650	0.990	1.390	2.100	2.780	3.560	4.790	6.530	9.740
3450	0.370	0.640	0.970	1.370	2.070	2.740	3.510	4.720	6.430	9.600
3500	0.360	0.630	0.960	1.350	2.040	2.710	3.460	4.650	6.340	9.460
3550	0.360	0.620	0.950	1.330	2.010	2.670	3.410	4.590	6.250	9.330
3600	0.350	0.610	0.930	1.310	1.980	2.630	3.370	4.530	6.170	9.200
3650	0.350	0.600	0.920	1.290	1.950	2.590	3.320	4.460	6.080	9.070
3700	0.340	0.600	0.910	1.280	1.930	2.560	3.280	4.400	6.000	8.950
3750	0.340	0.590	0.900	1.260	1.900	2.530	3.230	4.340	5.920	8.830
3800	0.330	0.580	0.880	1.240	1.880	2.490	3.190	4.290	5.840	8.720
3850	0.330	0.570	0.870	1.230	1.850	2.460	3.150	4.230	5.760	8.600
3900	0.330	0.570	0.860	1.210	1.830	2.430	3.110	4.180	5.690	8.490
3950	0.320	0.560	0.850	1.200	1.810	2.400	3.070	4.120	5.620	8.380
4000	0.320	0.550	0.840	1.180	1.780	2.370	3.030	4.070	5.550	8.280
4050	0.310	0.540	0.830	1.170	1.760	2.340	2.990	4.020	5.480	8.180
4100	0.310	0.540	0.820	1.150	1.740	2.310	2.960	3.970	5.410	8.080
4150	0.310	0.530	0.810	1.140	1.720	2.280	2.920	3.930	5.350	7.980
4200	0.300	0.530	0.800	1.130	1.700	2.250	2.890	3.880	5.280	7.890
4250	0.300	0.520	0.790	1.110	1.680	2.230	2.850	3.830	5.220	7.790
4300	0.300	0.510	0.780	1.100	1.660	2.200	2.820	3.790	5.160	7.700
4350	0.290	0.510	0.770	1.090	1.640	2.180	2.790	3.740	5.100	7.610
4400	0.290	0.500	0.760	1.070	1.620	2.150	2.750	3.700	5.040	7.530
4450	0.290	0.500	0.760	1.060	1.600	2.130	2.720	3.660	4.990	7.440
4500	0.280	0.490	0.750	1.050	1.590	2.100	2.690	3.620	4.930	7.360
4550	0.280	0.490	0.740	1.040	1.570	2.080	2.660	3.580	4.880	7.280
4600	0.280	0.480	0.730	1.030	1.550	2.060	2.630	3.540	4.830	7.200
4650	0.270	0.470	0.720	1.020	1.530	2.040	2.610	3.500	4.770	7.120
4700	0.270	0.470	0.710	1.010	1.520	2.010	2.580	3.470	4.720	7.050
4750	0.270	0.460	0.710	0.990	1.500	1.990	2.550	3.430	4.670	6.970
4800	0.260	0.460	0.700	0.980	1.490	1.970	2.520	3.390	4.620	6.900
4850	0.260	0.460	0.690	0.970	1.470	1.950	2.500	3.360	4.580	6.830
4900	0.260	0.450	0.690	0.960	1.460	1.930	2.470	3.320	4.530	6.760
4950	0.260	0.450	0.680	0.950	1.440	1.910	2.450	3.290	4.480	6.690
5000	0.250	0.440	0.670	0.950	1.430	1.890	2.420	3.260	4.440	6.620
Maximum Aspect Ratio	**7.3**	**6.8**	**6.5**	**6.3**	**5.9**	**4.9**	**4.3**	**3.8**	**3.3**	**2.9**

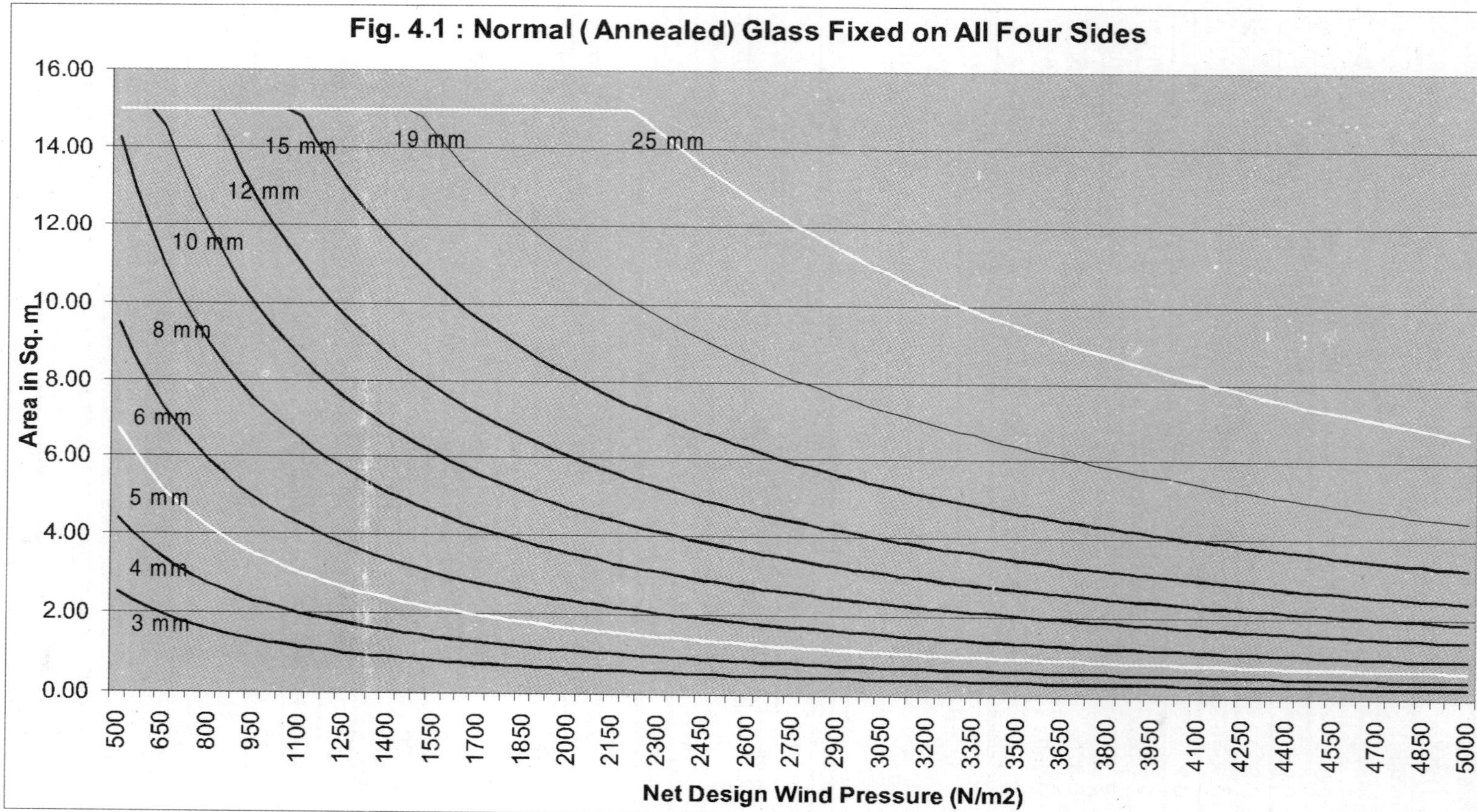
Area in Sq. m.
16.00
14.00
12.00
10.00
8.00
6.00
4.00
2.00
0.00
25 mm
19 mm
15 mm
12 mm
10 mm
8 mm
6 mm
5 mm
4 mm
3 mm
500
650
800
950
1100
1250
1400
1550
1700
1850
2000
2150
2300
2450
2600
2750
2900
3050
3200
3350
3500
3650
3800
3950
4100
4250
4400
4550
4700
4850
5000
Net Design Wind Pressure (N/m2)

Table 4.14 : Maximum Areas (m²) for Laminated Glass Fixed on All Four Sides.

Design Wind Pr. (N/m²)	Standard Nominal Thickness of Glass in mm					
	5.38	**6.38**	**8.38**	**10.38**	**12.38**	**16.38**
500	5.380	7.560	11.420	15.000	15.000	15.000
550	4.890	6.870	10.380	13.770	15.000	15.000
600	4.480	6.300	9.510	12.630	15.000	15.000
650	4.140	5.820	8.780	11.650	14.910	15.000
700	3.840	5.400	8.150	10.820	13.850	15.000
750	3.580	5.040	7.610	10.100	12.930	15.000
800	3.360	4.730	7.130	9.470	12.120	15.000
850	3.160	4.450	6.710	8.910	11.410	15.000
900	2.990	4.200	6.340	8.420	10.770	14.480
950	2.830	3.980	6.010	7.970	10.200	13.720
1000	2.690	3.780	5.710	7.580	9.690	13.030
1050	2.560	3.600	5.440	7.210	9.230	12.410
1100	2.440	3.440	5.190	6.890	8.810	11.850
1150	2.340	3.290	4.960	6.590	8.430	11.330
1200	2.240	3.150	4.760	6.310	8.080	10.860
1250	2.150	3.020	4.570	6.060	7.760	10.430
1300	2.070	2.910	4.390	5.830	7.460	10.020
1350	1.990	2.800	4.230	5.610	7.180	9.650
1400	1.920	2.700	4.080	5.410	6.920	9.310
1450	1.850	2.610	3.940	5.220	6.690	8.990
1500	1.790	2.520	3.810	5.050	6.460	8.690
1550	1.730	2.440	3.680	4.890	6.250	8.410
1600	1.680	2.360	3.570	4.730	6.060	8.150
1650	1.630	2.290	3.460	4.590	5.880	7.900
1700	1.580	2.220	3.360	4.460	5.700	7.670
1750	1.540	2.160	3.260	4.330	5.540	7.450
1800	1.490	2.100	3.170	4.210	5.390	7.240
1850	1.450	2.040	3.090	4.090	5.240	7.040
1900	1.410	1.990	3.000	3.990	5.100	6.860
1950	1.380	1.940	2.930	3.880	4.970	6.680
2000	1.340	1.890	2.850	3.790	4.850	6.520
2050	1.310	1.840	2.780	3.700	4.730	6.360
Maximum Aspect Ratio	**6.5**	**6.2**	**5.7**	**4.8**	**4.3**	**3.6**

	5.38	**6.38**	**8.38**	**10.38**	**12.38**	**16.38**
2100	1.280	1.800	2.720	3.610	4.620	6.210
2150	1.250	1.760	2.650	3.520	4.510	6.060
2200	1.220	1.720	2.590	3.440	4.410	5.920
2250	1.190	1.680	2.540	3.370	4.310	5.790
2300	1.170	1.640	2.480	3.290	4.220	5.670
2350	1.140	1.610	2.430	3.220	4.130	5.550
2400	1.120	1.580	2.380	3.160	4.040	5.430
2450	1.100	1.540	2.330	3.090	3.960	5.320
2500	1.080	1.510	2.280	3.030	3.880	5.210
2550	1.050	1.480	2.240	2.970	3.800	5.110
2600	1.030	1.450	2.200	2.910	3.730	5.010
2650	1.010	1.430	2.150	2.860	3.660	4.920
2700	1.000	1.400	2.110	2.810	3.590	4.830
2750	0.980	1.370	2.080	2.750	3.530	4.740
2800	0.960	1.350	2.040	2.710	3.460	4.650
2850	0.940	1.330	2.000	2.660	3.400	4.570
2900	0.930	1.300	1.970	2.610	3.340	4.490
2950	0.910	1.280	1.930	2.570	3.290	4.420
3000	0.900	1.260	1.900	2.530	3.230	4.340
3050	0.880	1.240	1.870	2.480	3.180	4.270
3100	0.870	1.220	1.840	2.440	3.130	4.200
3150	0.850	1.200	1.810	2.400	3.080	4.140
3200	0.840	1.180	1.780	2.370	3.030	4.070
3250	0.830	1.160	1.760	2.330	2.980	4.010
3300	0.810	1.150	1.730	2.300	2.940	3.950
3350	0.800	1.130	1.700	2.260	2.890	3.890
3400	0.790	1.110	1.680	2.230	2.850	3.830
3450	0.780	1.100	1.650	2.200	2.810	3.780
3500	0.770	1.080	1.630	2.160	2.770	3.720
3550	0.760	1.060	1.610	2.130	2.730	3.670
3600	0.750	1.050	1.590	2.100	2.690	3.620
Maximum Aspect Ratio	**6.5**	**6.2**	**5.7**	**4.8**	**4.3**	**3.6**

	5.38	**6.38**	**8.38**	**10.38**	**12.38**	**16.38**
3650	0.740	1.040	1.560	2.080	2.660	3.570
3700	0.730	1.020	1.540	2.050	2.620	3.520
3750	0.720	1.010	1.520	2.020	2.590	3.480
3800	0.710	0.990	1.500	1.990	2.550	3.430
3850	0.700	0.980	1.480	1.970	2.520	3.390
3900	0.690	0.970	1.460	1.940	2.490	3.340
3950	0.680	0.960	1.440	1.920	2.450	3.300
4000	0.670	0.950	1.430	1.890	2.420	3.260
4050	0.660	0.930	1.410	1.870	2.390	3.220
4100	0.660	0.920	1.390	1.850	2.360	3.180
4150	0.650	0.910	1.380	1.830	2.340	3.140
4200	0.640	0.900	1.360	1.800	2.310	3.100
4250	0.630	0.890	1.340	1.780	2.280	3.070
4300	0.630	0.880	1.330	1.760	2.250	3.030
4350	0.620	0.870	1.310	1.740	2.230	3.000
4400	0.610	0.860	1.300	1.720	2.200	2.960
4450	0.600	0.850	1.280	1.700	2.180	2.930
4500	0.600	0.840	1.270	1.680	2.150	2.900
4550	0.590	0.830	1.250	1.660	2.130	2.860
4600	0.580	0.820	1.240	1.650	2.110	2.830
4650	0.580	0.810	1.230	1.630	2.080	2.800
4700	0.570	0.800	1.210	1.610	2.060	2.770
4750	0.570	0.800	1.200	1.590	2.040	2.740
4800	0.560	0.790	1.190	1.580	2.020	2.720
4850	0.550	0.780	1.180	1.560	2.000	2.690
4900	0.550	0.770	1.160	1.550	1.980	2.660
4950	0.540	0.760	1.150	1.530	1.960	2.630
5000	0.540	0.760	1.140	1.520	1.940	2.610
Maximum Aspect Ratio	**6.5**	**6.2**	**5.7**	**4.8**	**4.3**	**3.6**

Fig. 4.2 : Laminated Glass Panel Fixed on All Four Sides

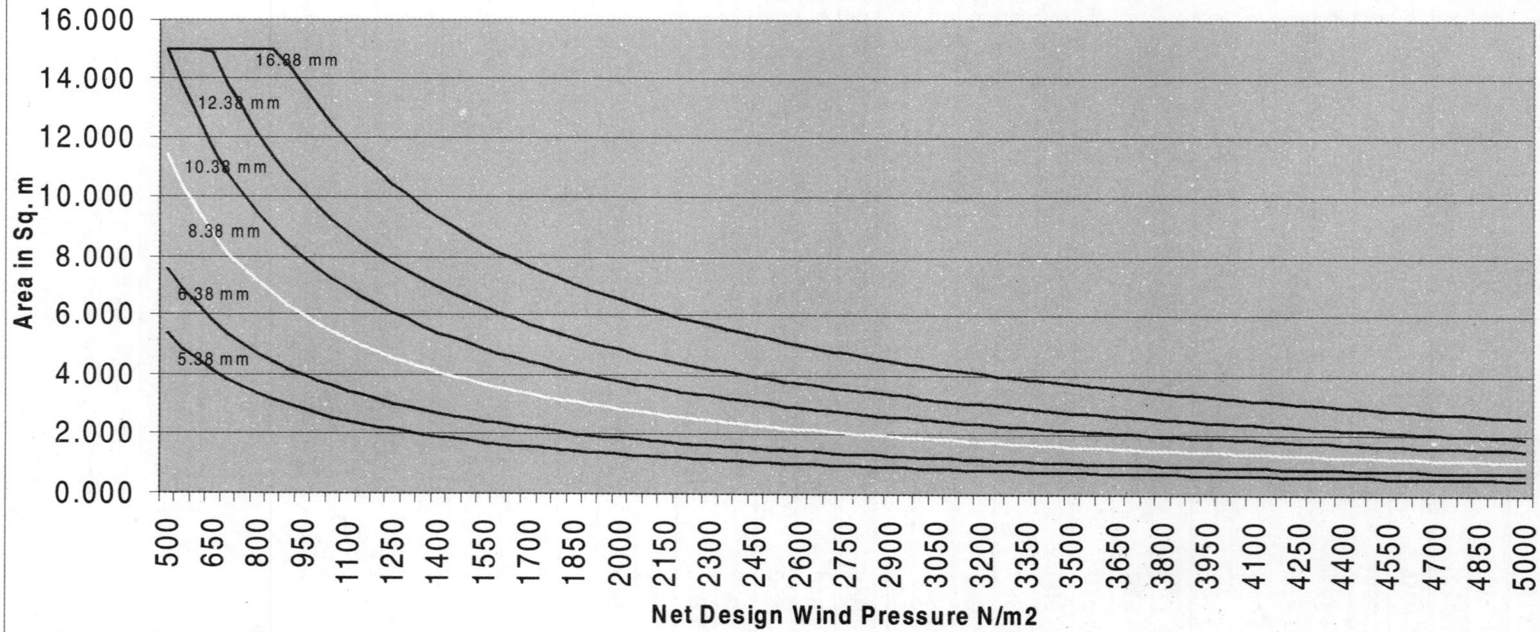

	3+3	4+4	5+5	6+6	8+8	10+10	12+12
3400	0.560	0.970	1.480	2.080	3.150	4.180	5.350
3450	0.550	0.960	1.460	2.050	3.100	4.120	5.270
3500	0.540	0.950	1.440	2.030	3.060	4.060	5.190
3550	0.540	0.930	1.420	2.000	3.010	4.000	5.120
3600	0.530	0.920	1.400	1.970	2.970	3.950	5.050
3650	0.520	0.910	1.380	1.940	2.930	3.890	4.980
3700	0.520	0.890	1.360	1.920	2.890	3.840	4.910
3750	0.510	0.880	1.340	1.890	2.850	3.790	4.850
3800	0.500	0.870	1.330	1.870	2.820	3.740	4.780
3850	0.500	0.860	1.310	1.840	2.780	3.690	4.720
3900	0.490	0.850	1.290	1.820	2.740	3.640	4.660
3950	0.480	0.840	1.280	1.790	2.710	3.600	4.600
4000	0.480	0.830	1.260	1.770	2.680	3.550	4.540
4050	0.470	0.820	1.240	1.750	2.640	3.510	4.490
4100	0.460	0.810	1.230	1.730	2.610	3.460	4.430
4150	0.460	0.800	1.210	1.710	2.580	3.420	4.380
4200	0.450	0.790	1.200	1.690	2.550	3.380	4.330
4250	0.450	0.780	1.190	1.670	2.520	3.340	4.280
4300	0.440	0.770	1.170	1.650	2.490	3.300	4.230
4350	0.440	0.760	1.160	1.630	2.460	3.270	4.180
4400	0.430	0.750	1.150	1.610	2.430	3.230	4.130
4450	0.430	0.740	1.130	1.590	2.400	3.190	4.080
4500	0.420	0.740	1.120	1.580	2.380	3.160	4.040
4550	0.420	0.730	1.110	1.560	2.350	3.120	4.000
4600	0.410	0.720	1.100	1.540	2.330	3.090	3.950
4650	0.410	0.710	1.080	1.520	2.300	3.050	3.910
4700	0.410	0.700	1.070	1.510	2.280	3.020	3.870
4750	0.400	0.700	1.060	1.490	2.250	2.990	3.830
4800	0.400	0.690	1.050	1.480	2.230	2.960	3.790
4850	0.390	0.680	1.040	1.460	2.210	2.930	3.750
4900	0.390	0.680	1.030	1.450	2.180	2.900	3.710
4950	0.390	0.670	1.020	1.430	2.160	2.870	3.670
5000	0.380	0.660	1.010	1.420	2.140	2.840	3.640
Maximum Aspect Ratio	**7.3**	**6.8**	**6.5**	**6.3**	**5.9**	**4.9**	**4.3**

Fig. 4.4 : Insulating Glass Fixed on All Four Sides

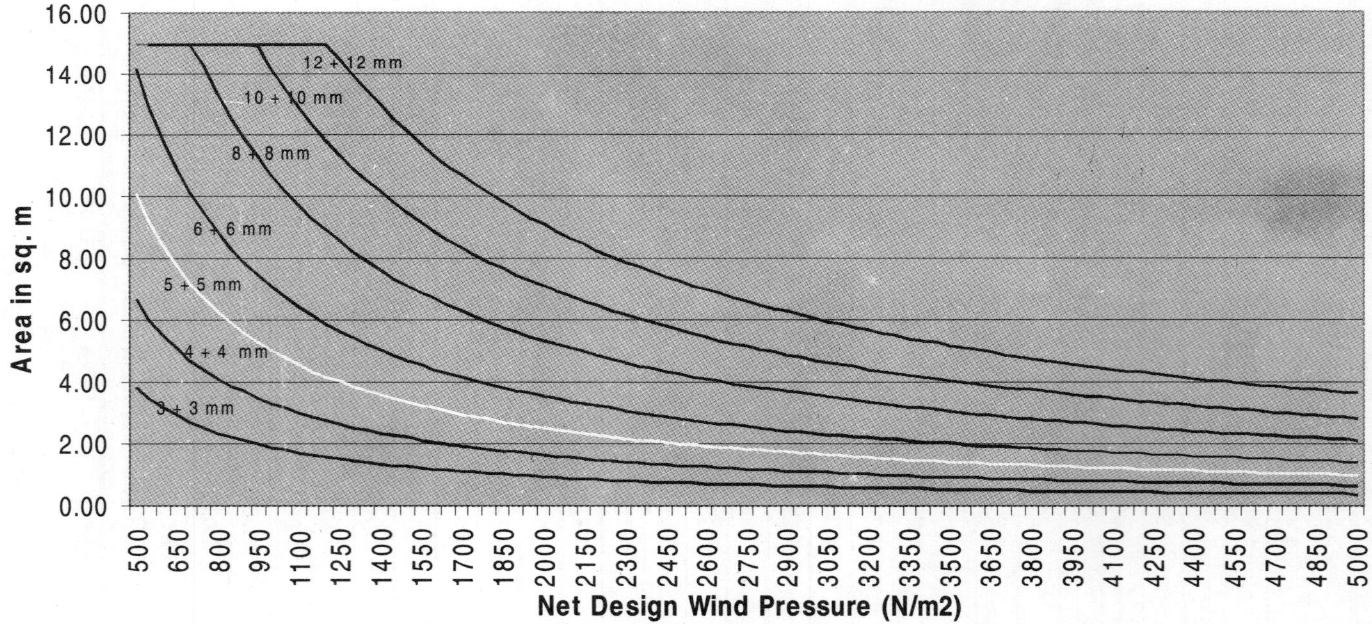

Table No. 4.17 : Maximum Span (m) for Normal Glass Fixed on Two Opposite Sides.

Design Wind Pr. (N/m²)	Standard Nominal Thickness of Glass in mm									
	3.0	**4.0**	**5.0**	**6.0**	**8.0**	**10.0**	**12.0**	**15.0**	**19.0**	**25.0**
500	0.59	0.78	0.98	1.18	1.51	1.89	2.26	2.83	3.59	4.00
550	0.56	0.75	0.93	1.12	1.44	1.8	2.16	2.7	3.42	4.00
600	0.54	0.72	0.89	1.07	1.38	1.72	2.07	2.58	3.27	4.00
650	0.52	0.69	0.86	1.03	1.32	1.66	1.99	2.48	3.14	4.00
700	0.50	0.66	0.83	0.99	1.28	1.6	1.91	2.39	3.03	3.99
750	0.48	0.64	0.80	0.96	1.23	1.54	1.85	2.31	2.93	3.85
800	0.46	0.62	0.77	0.93	1.19	1.49	1.79	2.24	2.83	3.73
850	0.45	0.60	0.75	0.90	1.16	1.45	1.74	2.17	2.75	3.62
900	0.44	0.58	0.73	0.88	1.13	1.41	1.69	2.11	2.67	3.52
950	0.43	0.57	0.71	0.85	1.10	1.37	1.64	2.05	2.6	3.42
1000	0.42	0.55	0.69	0.83	1.07	1.33	1.6	2	2.54	3.34
1050	0.41	0.54	0.68	0.81	1.04	1.3	1.56	1.95	2.47	3.26
1100	0.40	0.53	0.66	0.79	1.02	1.27	1.53	1.91	2.42	3.18
1150	0.39	0.52	0.65	0.77	1.00	1.24	1.49	1.87	2.36	3.11
1200	0.38	0.51	0.63	0.76	0.97	1.22	1.46	1.83	2.31	3.05
1250	0.37	0.5	0.62	0.74	0.95	1.19	1.43	1.79	2.27	2.98
1300	0.36	0.49	0.61	0.73	0.94	1.17	1.4	1.76	2.22	2.93
1350	0.36	0.48	0.6	0.72	0.92	1.15	1.38	1.72	2.18	2.87
1400	0.35	0.47	0.59	0.7	0.9	1.13	1.35	1.69	2.14	2.82
1450	0.35	0.46	0.58	0.69	0.89	1.11	1.33	1.66	2.11	2.77
1500	0.34	0.45	0.57	0.68	0.87	1.09	1.31	1.63	2.07	2.72
1550	0.33	0.45	0.56	0.67	0.86	1.07	1.29	1.61	2.04	2.68
1600	0.33	0.44	0.55	0.66	0.84	1.05	1.27	1.58	2	2.64
1650	0.32	0.43	0.54	0.65	0.83	1.04	1.25	1.56	1.97	2.6
1700	0.32	0.42	0.53	0.64	0.82	1.02	1.23	1.54	1.94	2.56
1750	0.31	0.42	0.52	0.63	0.81	1.01	1.21	1.51	1.92	2.52
1800	0.31	0.41	0.52	0.62	0.8	0.99	1.19	1.49	1.89	2.49
1850	0.31	0.41	0.51	0.61	0.78	0.98	1.18	1.47	1.86	2.45
1900	0.3	0.4	0.5	0.6	0.77	0.97	1.16	1.45	1.84	2.42

	3.0	4.0	5.0	6.0	8.0	10.0	12.0	15.0	19.0	25.0
1950	0.3	0.4	0.5	0.6	0.76	0.96	1.15	1.43	1.82	2.39
2000	0.29	0.39	0.49	0.59	0.75	0.94	1.13	1.42	1.79	2.36
2050	0.29	0.39	0.48	0.58	0.75	0.93	1.12	1.4	1.77	2.33
2100	0.29	0.38	0.48	0.57	0.74	0.92	1.11	1.38	1.75	2.3
2150	0.28	0.38	0.47	0.57	0.73	0.91	1.09	1.37	1.73	2.28
2200	0.28	0.37	0.47	0.56	0.72	0.9	1.08	1.35	1.71	2.25
2250	0.28	0.37	0.46	0.55	0.71	0.89	1.07	1.33	1.69	2.22
2300	0.27	0.37	0.46	0.55	0.7	0.88	1.06	1.32	1.67	2.2
2350	0.27	0.36	0.45	0.54	0.7	0.87	1.04	1.31	1.65	2.18
2400	0.27	0.36	0.45	0.54	0.69	0.86	1.03	1.29	1.64	2.15
2450	0.27	0.35	0.44	0.53	0.68	0.85	1.02	1.28	1.62	2.13
2500	0.26	0.35	0.44	0.53	0.68	0.84	1.01	1.27	1.6	2.11
2550	0.26	0.35	0.43	0.52	0.67	0.84	1	1.25	1.59	2.09
2600	0.26	0.34	0.43	0.52	0.66	0.83	0.99	1.24	1.57	2.07
2650	0.26	0.34	0.43	0.51	0.66	0.82	0.98	1.23	1.56	2.05
2700	0.25	0.34	0.42	0.51	0.65	0.81	0.97	1.22	1.54	2.03
2750	0.25	0.33	0.42	0.5	0.64	0.8	0.97	1.21	1.53	2.01
2800	0.25	0.33	0.41	0.5	0.64	0.8	0.96	1.2	1.52	1.99
2850	0.25	0.33	0.41	0.49	0.63	0.79	0.95	1.19	1.5	1.98
2900	0.24	0.33	0.41	0.49	0.63	0.78	0.94	1.18	1.49	1.96
2950	0.24	0.32	0.4	0.48	0.62	0.78	0.93	1.17	1.48	1.94
3000	0.24	0.32	0.4	0.48	0.62	0.77	0.92	1.16	1.46	1.93
3050	0.24	0.32	0.4	0.48	0.61	0.76	0.92	1.15	1.45	1.91
3100	0.24	0.31	0.39	0.47	0.61	0.76	0.91	1.14	1.44	1.89
3150	0.23	0.31	0.39	0.47	0.6	0.75	0.9	1.13	1.43	1.88
3200	0.23	0.31	0.39	0.46	0.6	0.75	0.9	1.12	1.42	1.86
3250	0.23	0.31	0.38	0.46	0.59	0.74	0.89	1.11	1.41	1.85
3300	0.23	0.3	0.38	0.46	0.59	0.73	0.88	1.1	1.4	1.84
3350	0.23	0.3	0.38	0.45	0.58	0.73	0.87	1.09	1.39	1.82
3400	0.23	0.3	0.38	0.45	0.58	0.72	0.87	1.09	1.38	1.81
3450	0.22	0.3	0.37	0.45	0.57	0.72	0.86	1.08	1.37	1.8
3500	0.22	0.3	0.37	0.44	0.57	0.71	0.86	1.07	1.36	1.78

	3.0	4.0	5.0	6.0	8.0	10.0	12.0	15.0	19.0	25.0
3550	0.22	0.29	0.37	0.44	0.57	0.71	0.85	1.06	1.35	1.77
3600	0.22	0.29	0.37	0.44	0.56	0.7	0.84	1.05	1.34	1.76
3650	0.22	0.29	0.36	0.43	0.56	0.7	0.84	1.05	1.33	1.75
3700	0.22	0.29	0.36	0.43	0.56	0.69	0.83	1.04	1.32	1.73
3750	0.21	0.29	0.36	0.43	0.55	0.69	0.83	1.03	1.31	1.72
3800	0.21	0.28	0.36	0.43	0.55	0.68	0.82	1.03	1.3	1.71
3850	0.21	0.28	0.35	0.42	0.54	0.68	0.82	1.02	1.29	1.7
3900	0.21	0.28	0.35	0.42	0.54	0.68	0.81	1.01	1.28	1.69
3950	0.21	0.28	0.35	0.42	0.54	0.67	0.81	1.01	1.28	1.68
4000	0.21	0.28	0.35	0.42	0.53	0.67	0.8	1	1.27	1.67
4050	0.21	0.28	0.34	0.41	0.53	0.66	0.8	0.99	1.26	1.66
4100	0.21	0.27	0.34	0.41	0.53	0.66	0.79	0.99	1.25	1.65
4150	0.2	0.27	0.34	0.41	0.52	0.66	0.79	0.98	1.24	1.64
4200	0.2	0.27	0.34	0.41	0.52	0.65	0.78	0.98	1.24	1.63
4250	0.2	0.27	0.34	0.4	0.52	0.65	0.78	0.97	1.23	1.62
4300	0.2	0.27	0.33	0.4	0.51	0.64	0.77	0.97	1.22	1.61
4350	0.2	0.27	0.33	0.4	0.51	0.64	0.77	0.96	1.22	1.6
4400	0.2	0.26	0.33	0.4	0.51	0.64	0.76	0.95	1.21	1.59
4450	0.2	0.26	0.33	0.39	0.51	0.63	0.76	0.95	1.2	1.58
4500	0.2	0.26	0.33	0.39	0.5	0.63	0.75	0.94	1.2	1.57
4550	0.19	0.26	0.32	0.39	0.5	0.63	0.75	0.94	1.19	1.56
4600	0.19	0.26	0.32	0.39	0.5	0.62	0.75	0.93	1.18	1.56
4650	0.19	0.26	0.32	0.39	0.5	0.62	0.74	0.93	1.18	1.55
4700	0.19	0.26	0.32	0.38	0.49	0.62	0.74	0.92	1.17	1.54
4750	0.19	0.25	0.32	0.38	0.49	0.61	0.73	0.92	1.16	1.53
4800	0.19	0.25	0.32	0.38	0.49	0.61	0.73	0.91	1.16	1.52
4850	0.19	0.25	0.31	0.38	0.48	0.61	0.73	0.91	1.15	1.51
4900	0.19	0.25	0.31	0.38	0.48	0.6	0.72	0.9	1.15	1.51
4950	0.19	0.25	0.31	0.37	0.48	0.6	0.72	0.9	1.14	1.5
5000	0.19	0.25	0.31	0.37	0.48	0.6	0.72	0.9	1.13	1.49

Fig. 4.5 Normal Glass Fixed on Two Opposite Sides.

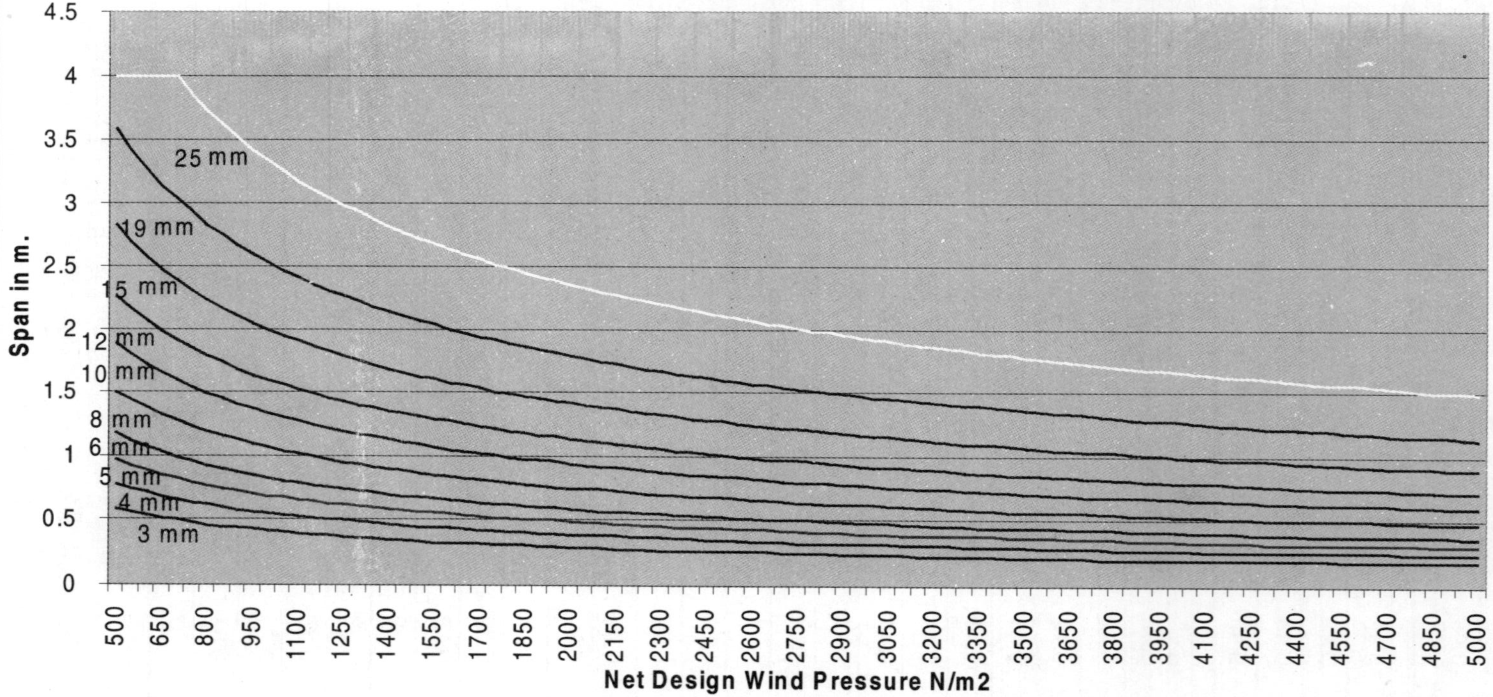

Table No. 4.18 : Maximum Span (m) for Laminated Glass Fixed on Two Opposite Sides.

Design Wind Pr. (N/m²)	Standard Nominal Thickness of Glass in mm					
	5.38	**6.38**	**8.38**	**10.38**	**12.38**	**16.38**
500	0.88	1.05	1.35	1.69	2.03	2.53
550	0.84	1.00	1.29	1.61	1.93	2.41
600	0.80	0.96	1.23	1.54	1.85	2.31
650	0.77	0.92	1.18	1.48	1.78	2.22
700	0.74	0.89	1.14	1.43	1.71	2.14
750	0.72	0.86	1.10	1.38	1.65	2.07
800	0.69	0.83	1.07	1.33	1.60	2.00
850	0.67	0.81	1.04	1.29	1.55	1.94
900	0.65	0.78	1.01	1.26	1.51	1.89
950	0.64	0.76	0.98	1.22	1.47	1.84
1000	0.62	0.74	0.95	1.19	1.43	1.79
1050	0.60	0.73	0.93	1.16	1.40	1.75
1100	0.59	0.71	0.91	1.14	1.37	1.71
1150	0.58	0.69	0.89	1.11	1.34	1.67
1200	0.57	0.68	0.87	1.09	1.31	1.63
1250	0.55	0.66	0.85	1.07	1.28	1.60
1300	0.54	0.65	0.84	1.05	1.26	1.57
1350	0.53	0.64	0.82	1.03	1.23	1.54
1400	0.52	0.63	0.81	1.01	1.21	1.51
1450	0.51	0.62	0.79	0.99	1.19	1.49
1500	0.51	0.61	0.78	0.97	1.17	1.46
1550	0.50	0.60	0.77	0.96	1.15	1.44
1600	0.49	0.59	0.75	0.94	1.13	1.42
1650	0.48	0.58	0.74	0.93	1.12	1.39
1700	0.48	0.57	0.73	0.92	1.10	1.37
1750	0.47	0.56	0.72	0.9	1.08	1.35
1800	0.46	0.55	0.71	0.89	1.07	1.33
1850	0.46	0.55	0.70	0.88	1.05	1.32
1900	0.45	0.54	0.69	0.87	1.04	1.30
1950	0.44	0.53	0.68	0.85	1.03	1.28

	5.38	**6.38**	**8.38**	**10.38**	**12.38**	**16.38**
2000	0.44	0.53	0.68	0.84	1.01	1.27
2050	0.43	0.52	0.67	0.83	1.00	1.25
2100	0.43	0.51	0.66	0.82	0.99	1.24
2150	0.42	0.51	0.65	0.81	0.98	1.22
2200	0.42	0.50	0.64	0.80	0.97	1.21
2250	0.41	0.50	0.64	0.80	0.95	1.19
2300	0.41	0.49	0.63	0.79	0.94	1.18
2350	0.40	0.48	0.62	0.78	0.93	1.17
2400	0.40	0.48	0.62	0.77	0.92	1.16
2450	0.40	0.47	0.61	0.76	0.92	1.14
2500	0.39	0.47	0.60	0.75	0.91	1.13
2550	0.39	0.47	0.60	0.75	0.90	1.12
2600	0.38	0.46	0.59	0.74	0.89	1.11
2650	0.38	0.46	0.59	0.73	0.88	1.10
2700	0.38	0.45	0.58	0.73	0.87	1.09
2750	0.37	0.45	0.58	0.72	0.86	1.08
2800	0.37	0.44	0.57	0.71	0.86	1.07
2850	0.37	0.44	0.57	0.71	0.85	1.06
2900	0.36	0.44	0.56	0.7	0.84	1.05
2950	0.36	0.43	0.56	0.69	0.83	1.04
3000	0.36	0.43	0.55	0.69	0.83	1.03
3050	0.35	0.43	0.55	0.68	0.82	1.03
3100	0.35	0.42	0.54	0.68	0.81	1.02
3150	0.35	0.42	0.54	0.67	0.81	1.01
3200	0.35	0.42	0.53	0.67	0.80	1.00
3250	0.34	0.41	0.53	0.66	0.79	0.99
3300	0.34	0.41	0.53	0.66	0.79	0.99
3350	0.34	0.41	0.52	0.65	0.78	0.98
3400	0.34	0.40	0.52	0.65	0.78	0.97
3450	0.33	0.40	0.51	0.64	0.77	0.96
3500	0.33	0.40	0.51	0.64	0.77	0.96
3550	0.33	0.39	0.51	0.63	0.76	0.95
3600	0.33	0.39	0.50	0.63	0.75	0.94

	5.38	6.38	8.38	10.38	12.38	16.38
3650	0.32	0.39	0.50	0.62	0.75	0.94
3700	0.32	0.39	0.50	0.62	0.74	0.93
3750	0.32	0.38	0.49	0.62	0.74	0.92
3800	0.32	0.38	0.49	0.61	0.73	0.92
3850	0.32	0.38	0.49	0.61	0.73	0.91
3900	0.31	0.38	0.48	0.6	0.73	0.91
3950	0.31	0.37	0.48	0.6	0.72	0.90
4000	0.31	0.37	0.48	0.6	0.72	0.90
4050	0.31	0.37	0.47	0.59	0.71	0.89
4100	0.31	0.37	0.47	0.59	0.71	0.88
4150	0.30	0.36	0.47	0.59	0.70	0.88
4200	0.30	0.36	0.47	0.58	0.70	0.87
4250	0.30	0.36	0.46	0.58	0.69	0.87
4300	0.30	0.36	0.46	0.58	0.69	0.86
4350	0.30	0.36	0.46	0.57	0.69	0.86
4400	0.30	0.35	0.46	0.57	0.68	0.85
4450	0.29	0.35	0.45	0.57	0.68	0.85
4500	0.29	0.35	0.45	0.56	0.68	0.84
4550	0.29	0.35	0.45	0.56	0.67	0.84
4600	0.29	0.35	0.45	0.56	0.67	0.83
4650	0.29	0.34	0.44	0.55	0.66	0.83
4700	0.29	0.34	0.44	0.55	0.66	0.83
4750	0.28	0.34	0.44	0.55	0.66	0.82
4800	0.28	0.34	0.44	0.54	0.65	0.82
4850	0.28	0.34	0.43	0.54	0.65	0.81
4900	0.28	0.34	0.43	0.54	0.65	0.81
4950	0.28	0.33	0.43	0.54	0.64	0.80
5000	0.28	0.33	0.43	0.53	0.64	0.80

Fig. 4.6 : Laminated Glass Fixed on Two Opposite Sides

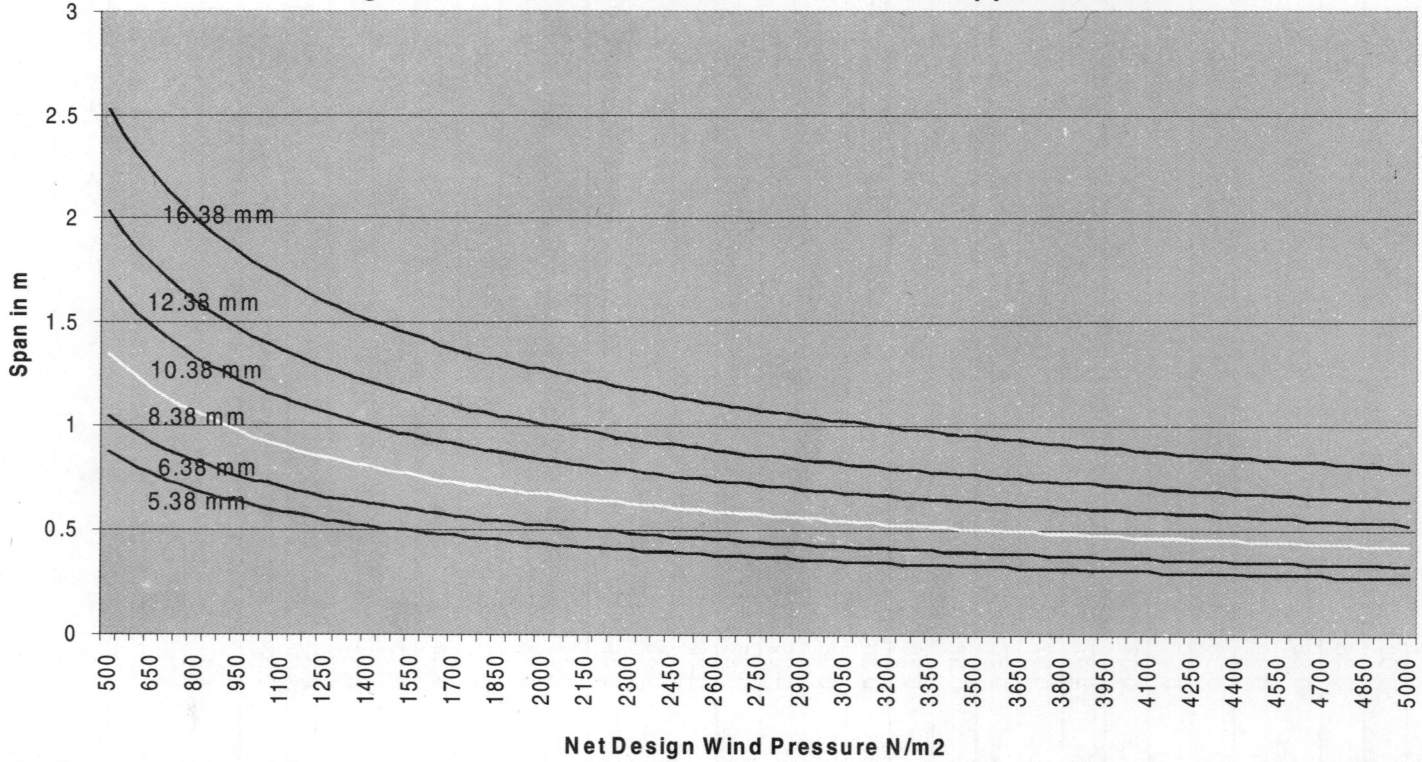

Table No. 4.19 : Maximum Span (m) for Tempered Glass Fixed on Two Opposite Sides.

Design Wind Pr. (N/m²)	Standard Nominal Thickness of Glass in mm									
	3.0	**4.0**	**5.0**	**6.0**	**8.0**	**10.0**	**12.0**	**15.0**	**19.0**	**25.0**
500	0.69	0.92	1.16	1.39	1.84	2.3	2.76	3.45	4.00	4.00
550	0.66	0.88	1.1	1.32	1.75	2.19	2.63	3.29	4.00	4.00
600	0.63	0.84	1.05	1.27	1.68	2.10	2.52	3.15	3.99	4.00
650	0.61	0.81	1.01	1.22	1.61	2.02	2.42	3.02	3.83	4.00
700	0.59	0.78	0.98	1.17	1.55	1.94	2.33	2.91	3.69	4.00
750	0.57	0.75	0.94	1.13	1.50	1.88	2.25	2.81	3.57	4.00
800	0.55	0.73	0.91	1.10	1.45	1.82	2.18	2.73	3.45	4.00
850	0.53	0.71	0.89	1.06	1.41	1.76	2.12	2.64	3.35	4.00
900	0.52	0.69	0.86	1.03	1.37	1.71	2.06	2.57	3.25	4.00
950	0.50	0.67	0.84	1.01	1.33	1.67	2.00	2.50	3.17	4.00
1000	0.49	0.65	0.82	0.98	1.30	1.63	1.95	2.44	3.09	4.00
1050	0.48	0.64	0.80	0.96	1.27	1.59	1.90	2.38	3.01	3.96
1100	0.47	0.62	0.78	0.94	1.24	1.55	1.86	2.32	2.94	3.87
1150	0.46	0.61	0.76	0.91	1.21	1.52	1.82	2.27	2.88	3.79
1200	0.45	0.60	0.75	0.90	1.19	1.48	1.78	2.23	2.82	3.71
1250	0.44	0.58	0.73	0.88	1.16	1.45	1.74	2.18	2.76	3.63
1300	0.43	0.57	0.72	0.86	1.14	1.43	1.71	2.14	2.71	3.56
1350	0.42	0.56	0.70	0.84	1.12	1.40	1.68	2.10	2.66	3.50
1400	0.41	0.55	0.69	0.83	1.10	1.37	1.65	2.06	2.61	3.43
1450	0.41	0.54	0.68	0.81	1.08	1.35	1.62	2.02	2.56	3.37
1500	0.40	0.53	0.67	0.80	1.06	1.33	1.59	1.99	2.52	3.32
1550	0.39	0.53	0.66	0.79	1.04	1.31	1.57	1.96	2.48	3.26
1600	0.39	0.52	0.65	0.78	1.03	1.28	1.54	1.93	2.44	3.21
1650	0.38	0.51	0.64	0.76	1.01	1.27	1.52	1.90	2.40	3.16
1700	0.38	0.50	0.63	0.75	1.00	1.25	1.50	1.87	2.37	3.12
1750	0.37	0.49	0.62	0.74	0.98	1.23	1.47	1.84	2.33	3.07
1800	0.37	0.49	0.61	0.73	0.97	1.21	1.45	1.82	2.30	3.03
1850	0.36	0.48	0.60	0.72	0.96	1.19	1.43	1.79	2.27	2.99
1900	0.36	0.47	0.59	0.71	0.94	1.18	1.41	1.77	2.24	2.95
1950	0.35	0.47	0.59	0.70	0.93	1.16	1.40	1.75	2.21	2.91

	3.0	4.0	5.0	6.0	8.0	10.0	12.0	15.0	19.0	25.0
2000	0.35	0.46	0.58	0.69	0.92	1.15	1.38	1.72	2.18	2.87
2050	0.34	0.46	0.57	0.68	0.91	1.13	1.36	1.70	2.16	2.84
2100	0.34	0.45	0.56	0.68	0.90	1.12	1.35	1.68	2.13	2.80
2150	0.33	0.45	0.56	0.67	0.89	1.11	1.33	1.66	2.11	2.77
2200	0.33	0.44	0.55	0.66	0.88	1.10	1.31	1.64	2.08	2.74
2250	0.33	0.44	0.54	0.65	0.87	1.08	1.30	1.63	2.06	2.71
2300	0.32	0.43	0.54	0.65	0.86	1.07	1.29	1.61	2.04	2.68
2350	0.32	0.43	0.53	0.64	0.85	1.06	1.27	1.59	2.01	2.65
2400	0.32	0.42	0.53	0.63	0.84	1.05	1.26	1.57	1.99	2.62
2450	0.31	0.42	0.52	0.63	0.83	1.04	1.25	1.56	1.97	2.60
2500	0.31	0.41	0.52	0.62	0.82	1.03	1.23	1.54	1.95	2.57
2550	0.31	0.41	0.51	0.61	0.81	1.02	1.22	1.53	1.93	2.54
2600	0.30	0.41	0.51	0.61	0.81	1.01	1.21	1.51	1.91	2.52
2650	0.30	0.40	0.50	0.60	0.80	1.00	1.20	1.50	1.90	2.50
2700	0.30	0.40	0.50	0.60	0.79	0.99	1.19	1.48	1.88	2.47
2750	0.30	0.39	0.49	0.59	0.78	0.98	1.18	1.47	1.86	2.45
2800	0.29	0.39	0.49	0.59	0.78	0.97	1.17	1.46	1.85	2.43
2850	0.29	0.39	0.48	0.58	0.77	0.96	1.16	1.44	1.83	2.41
2900	0.29	0.38	0.48	0.58	0.76	0.95	1.15	1.43	1.81	2.39
2950	0.29	0.38	0.48	0.57	0.76	0.95	1.14	1.42	1.8	2.37
3000	0.28	0.38	0.47	0.57	0.75	0.94	1.13	1.41	1.78	2.35
3050	0.28	0.37	0.47	0.56	0.74	0.93	1.12	1.40	1.77	2.33
3100	0.28	0.37	0.46	0.56	0.74	0.92	1.11	1.38	1.75	2.31
3150	0.28	0.37	0.46	0.55	0.73	0.92	1.1	1.37	1.74	2.29
3200	0.27	0.37	0.46	0.55	0.73	0.91	1.09	1.36	1.73	2.27
3250	0.27	0.36	0.45	0.54	0.72	0.9	1.08	1.35	1.71	2.25
3300	0.27	0.36	0.45	0.54	0.72	0.89	1.07	1.34	1.7	2.24
3350	0.27	0.36	0.45	0.54	0.71	0.89	1.07	1.33	1.69	2.22
3400	0.27	0.35	0.44	0.53	0.71	0.88	1.06	1.32	1.67	2.20
3450	0.26	0.35	0.44	0.53	0.70	0.87	1.05	1.31	1.66	2.19
3500	0.26	0.35	0.44	0.52	0.69	0.87	1.04	1.30	1.65	2.17
3550	0.26	0.35	0.43	0.52	0.69	0.86	1.03	1.29	1.64	2.16
3600	0.26	0.34	0.43	0.52	0.69	0.86	1.03	1.28	1.63	2.14

	3.0	4.0	5.0	6.0	8.0	10.0	12.0	15.0	19.0	25.0
3650	0.26	0.34	0.43	0.51	0.68	0.85	1.02	1.28	1.62	2.13
3700	0.25	0.34	0.42	0.51	0.68	0.84	1.01	1.27	1.61	2.11
3750	0.25	0.34	0.42	0.51	0.67	0.84	1.01	1.26	1.59	2.1
3800	0.25	0.34	0.42	0.50	0.67	0.83	1.00	1.25	1.58	2.08
3850	0.25	0.33	0.42	0.50	0.66	0.83	0.99	1.24	1.57	2.07
3900	0.25	0.33	0.41	0.50	0.66	0.82	0.99	1.23	1.56	2.06
3950	0.25	0.33	0.41	0.49	0.65	0.82	0.98	1.23	1.55	2.04
4000	0.25	0.33	0.41	0.49	0.65	0.81	0.98	1.22	1.54	2.03
4050	0.24	0.32	0.41	0.49	0.65	0.81	0.97	1.21	1.53	2.02
4100	0.24	0.32	0.40	0.48	0.64	0.80	0.96	1.20	1.52	2.01
4150	0.24	0.32	0.40	0.48	0.64	0.80	0.96	1.20	1.52	1.99
4200	0.24	0.32	0.40	0.48	0.63	0.79	0.95	1.19	1.51	1.98
4250	0.24	0.32	0.40	0.48	0.63	0.79	0.95	1.18	1.5	1.97
4300	0.24	0.32	0.39	0.47	0.63	0.78	0.94	1.18	1.49	1.96
4350	0.24	0.31	0.39	0.47	0.62	0.78	0.93	1.17	1.48	1.95
4400	0.23	0.31	0.39	0.47	0.62	0.77	0.93	1.16	1.47	1.94
4450	0.23	0.31	0.39	0.46	0.62	0.77	0.92	1.16	1.46	1.93
4500	0.23	0.31	0.39	0.46	0.61	0.77	0.92	1.15	1.46	1.92
4550	0.23	0.31	0.38	0.46	0.61	0.76	0.91	1.14	1.45	1.90
4600	0.23	0.30	0.38	0.46	0.61	0.76	0.91	1.14	1.44	1.89
4650	0.23	0.30	0.38	0.45	0.60	0.75	0.90	1.13	1.43	1.88
4700	0.23	0.30	0.38	0.45	0.60	0.75	0.90	1.12	1.42	1.87
4750	0.22	0.30	0.37	0.45	0.60	0.75	0.89	1.12	1.42	1.86
4800	0.22	0.30	0.37	0.45	0.59	0.74	0.89	1.11	1.41	1.85
4850	0.22	0.30	0.37	0.45	0.59	0.74	0.89	1.11	1.40	1.84
4900	0.22	0.30	0.37	0.44	0.59	0.73	0.88	1.10	1.39	1.84
4950	0.22	0.29	0.37	0.44	0.58	0.73	0.88	1.10	1.39	1.83
5000	0.22	0.29	0.37	0.44	0.58	0.73	0.87	1.09	1.38	1.82

Fig. 4.7 : Tempered Glass Supported on Two Opposites Sides

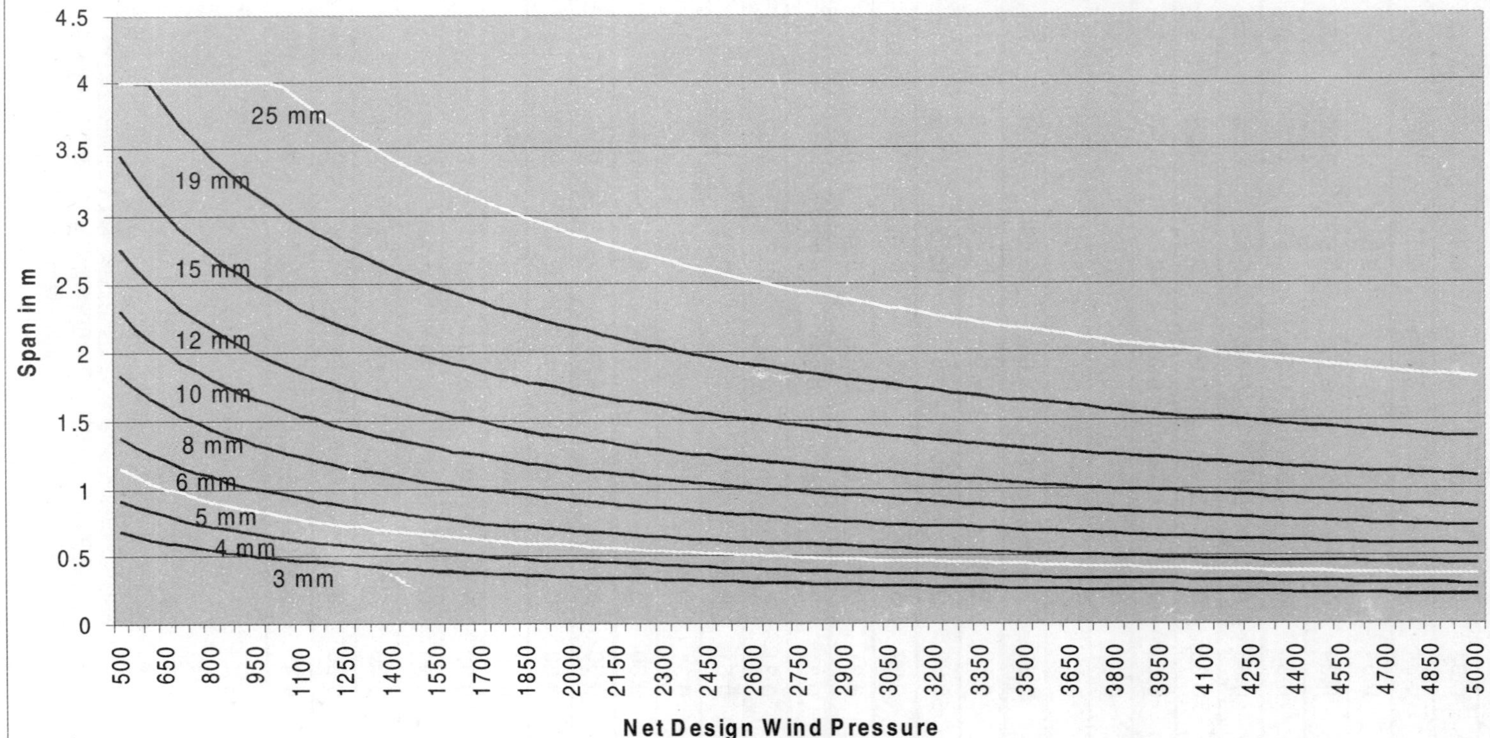

such as beads, clips or bolt fixings. Structural glazing with sealants allows perfectly uniform large glazed surfaces, not interrupted by traditional frames or any other supporting or fitting system projecting out of the frame. Instead of being fitted in a frame, the glass is fixed to a support, which in turn is attached to a structural element of the building, the tightness of the whole system being obtained by a silicone seal. The glass is fixed on its support by means of a silicone seal along the edges of the internal surface (fig. 5.4). This technique can be used with almost all types of glass, including insulating glass units.

The glazing is prepared in a factory by mounting a structural seal support frame onto the glass, complete with appropriate setting blocks, location blocks and distance pieces. On site, the support frame is attached to the building structure by mechanical means and the gap between the glazing is sealed.

5.3 BOLTED GLAZING

The bolted structural system is the less obtrusive, alternative to structural glazing, which holds the glass by means of visible metal parts and covers a small part of the glass surface. In some cases, the fixing holes can be drilled and countersunk, so that the bolts are embedded in the thickness of the glass it self.

For small or medium glazing not exceeding 7m in height and 50m in length, rigid bolted system can be used. This system is often used for shop front at street level. It consists of all glass assemblies which may include one or more single or double doors, transoms, fixed side panes, entirely made of tempered glass, assembled by various metal fittings and stiffened by mullions also made of tempered glass. This system consists of attaching the glass panels and the structure with rigid bolts and steel plates (fig. 5.5).

Bolted systems also use knuckled bolts for fixing of the glass, instead of the common fixed bolt. Knuckled bolt allow the glass, submitted to wind pressure, to be flexible and make a continuous curve instead of a double curvature. It limits sensibly the tension in the glass at suspension points. The new bolt, which is completely sealed, was intended

specially for roofs, but can be used as well in the elevations and has a special coating protecting it from wear and tear. The same coating also ensures the bolts are self-lubricant. On the inside there are two rings, which protect it from wind and water retention inside the bolt components to keep it flexible for life.

Bolted glass systems are often used on main entrance elevations of the big buildings, so that one can have unhindered view into the building. This technique even enables architects to make curved facades with flat panes.

5.4 FIN SUPPORTED GLAZING

Glass fins are used to achieve the 'all glass' clarity required by designers while meeting the structural requirements of the glazing system. It is important that the principles of design and installation are compiled with. A glass fin replaces a frame or mullion and must be assessed as to size and thickness and securely fixed or supported at the head and sill.

Any loading applied to the glass façade is transferred to the fin and then to the top and bottom fin shoes by way of a reaction load. The fins must be adhered to the façade glass with silicone sealant to cope with positive and negative loads. Clear silicone is commonly used but black silicone hides any minor bubbles and gives a better joint particularly with tinted and reflective glasses.

Fin systems are also designed with bolted joints. Structural glass is secured to a support structure by a variety of fittings, which are designed to meet the requirements of the structure. These fixings absorb force when the glass flexes under load and provide a secure connection between the glass component and the support structure. Fin systems create greater visibility in facades and increase levels of natural light in interiors.

5.5 CABLE STAYED GLAZING

Cable support systems like fin support system are also increasingly used. In these systems stainless steel or carbon fibre high strength cables are used to transfer loads to the main structure. Cables systems create greater levels of openness, visibility & natural light and enhance the dramatic appearance of the building (fig. 5.6).

5.6 SUSPENDED GLAZING

This method of providing a frameless glazing facade is to fix together a matrix of toughened glass lites, hung from

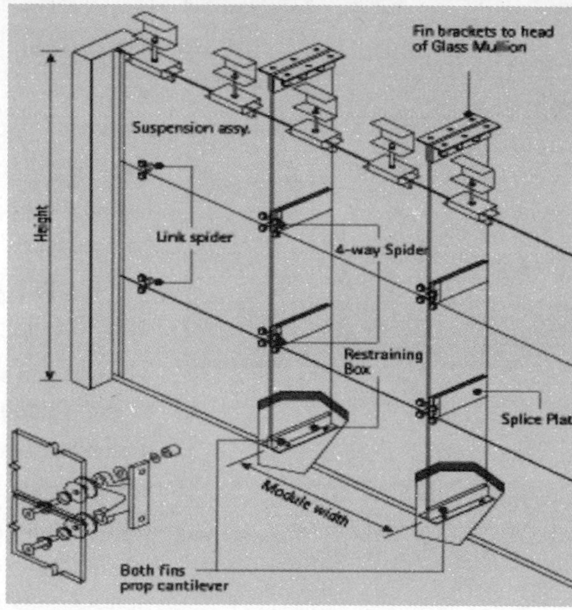

the building structure and allow designers to glaze large openings in buildings, without using metal frames or mullions, to create light and space with minimum visual barriers. Figure 5.7 illustrates a suspended glass assembly.

The glass façade is hung from the building structure like a curtain. The top tier panels are connected to the structure by adjustable hanger brackets and subsequent lower panels are connected by special fittings at their corners. The façade is located into channels at the perimeter and all glass joints and channels are sealed with silicone sealant. The hanging assembly is normally stabilized against wind load by glass fins located and fixed to the support structure with fittings at the joints.

This system is mainly used for tall glass panes, avoids flexing or buckling, which may happen if the panes would rest on their bottom edge. It also absorbs important movements of the building and it frees the lower frame from the weight of the glass.

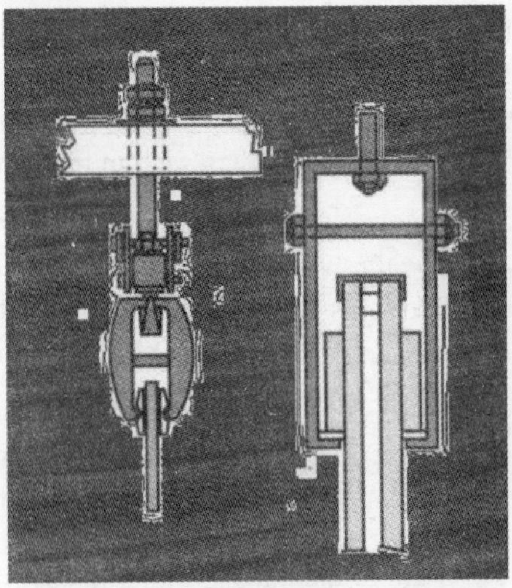

Monolithic glass panes are suspended by means of tongs, which press on both sides of the glass. Double glazed units, which cannot sustain pressure, are suspended by means of hooks. The system comprises a series of specially processed and toughened glass lites bolted together at their corners by small metal patch fittings. Pane-to-pane joints are sealed with a silicone building sealant, and toughened glass stabilizers are used at each vertical joint to provide lateral stiffness against wind loading. The assembly is suspended from the building structure by hangers bolted to its top edge and is sealed to the building in peripheral channels by neoprene strips or non-setting mastic.

The concept of the design ensures that the facade is, at all times, "floating" in the peripheral channeling, and problems, which might arise due to the differential movement between components, are eliminated. Assemblies, therefore, can be used to advantage when the design is to account for vibratory forces. Weather sealing is carried out at all joints in the facade using a structural silicone building sealant.

5.7 MIRROR INSTALLATION

- If the mirror is to be installed in high humidity area like bathrooms, then the edges should be sealed with any neutral cure silicone (like the Winsil 20 of GE or equivalent) because the silicone becomes a barrier for moisture and prevents the moisture coming in contact with the mirror backing thus increasing the life of the mirror.
- Only neutral silicones or special sealants recommended for fixing mirrors should be used as adhesive where necessary. However, as far as possible, mirrors should not be pasted on the wall/plywood.
- Any adhesive containing acid like acetoxy silicones will damage the mirror. Adhesives like Fevicol or Araldite should never be used, as their strong chemicals attack the mirror backing paint thereby resulting in black spots.

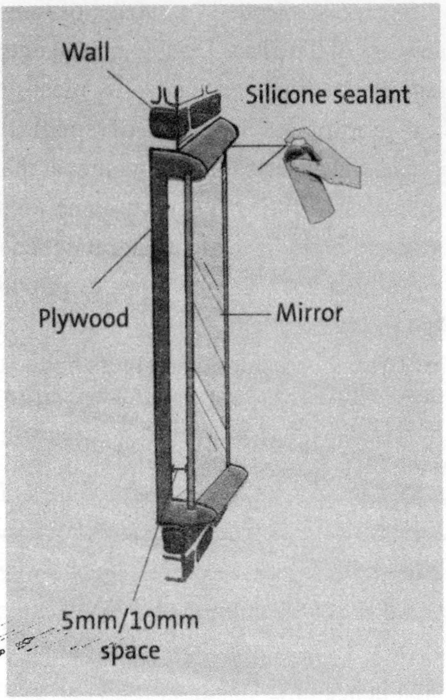

Instead 3M tapes can also be used for fixing mirrors to the plywood. These double-sided tapes also form a barrier to moisture thus preventing the mirror in coming in contact with the moisture that may seep in from the sides.

- The surface of the wall/plywood on which mirror is installed should be dry, as wet surface either due to seepage, pipe leakage or humidity can damage the mirror.
- If mirror is to be fixed to a freshly built surface like newly plastered wall or over Plaster of Paris, care should be taken that the surface is properly cured, as any amount of moisture or dampness will damage the mirror.
- Between the mirror and the mounting surface (wall, plywood etc.) a gap of 6mm to 10mm (for mirror of size more then 1 square meter) should be provided to ensure ventilation against humidity. The edges should be properly sealed as mentioned above.
- When adhering mirror to a surface, flatness is to be assured to avoid distortion.

5.8 ACCESSORIES FOR GLAZING SYSTEMS

Various types of accessories are required for different types of glazing systems to be used in buildings. The important ones are discussed below.

5.8.1 Aluminium Extrusions

All curtain wall system extrusions should be as per Indian standards 733-1983 and 1285 –1975. These extrusions are designed to resist seismic/wind loads in compliance with the requirements pertaining to the height of building.

These extrusions are anodized to prevent corrosion and to improve aesthetics

5.8.2 Mullions

Mullions are provided basically to transfer the dead load of the Curtain wall. Mullion joint is preferred at the bracket

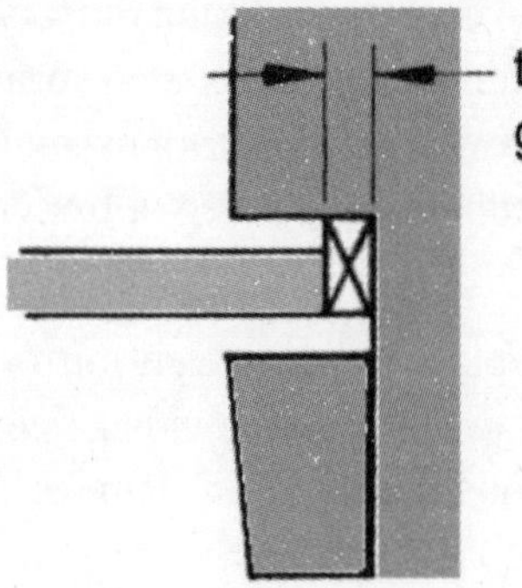

Location blocks

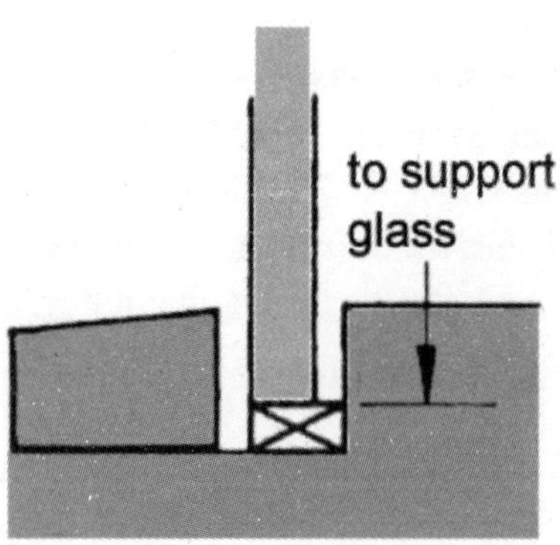

Setting blocks

location and aluminium stiffeners are provided for extra strength.

5.8.3 Transoms

Transoms are aluminium sections provided in between the mullions horizontally. They are Designed after due consideration of the floor height and the false-ceiling height.

5.8.4 Silicone Sealants

Silicone sealants are used to prevent the passage of moisture, air, dust and heat through al the joints. The putties formulated from linseed and other organic oils were the first generation sealants that perform satisfactory for 5 to 7 years. Mastics formulated from butyl and other synthetic oils, were improvement over putties. Their greater flexibility and movement permitted the use of aluminium and vinyl frames in place of wood.

Silicone represents the third generation sealants, which are derived from sand, and hence have outstanding resistance against sunlight, ozone, many alkalis and acids. They have excellent cohesive and adhesive strength. Silicones are used in glazing (glass-to-glass, metal-to-glass and metal-to-sealing), window sealing (aluminium/wood-to-concrete), and structural glazing applications.

5.8.5 Ethylene Propylene Diene Methylene (EPDM) Gaskets

Ethylene Propylene Diene Methylene (EPDM) gaskets acts as a line of defence for exterior seal. The purpose of these gaskets is to prevent water and air entering the building through the curtain wall. These extruded gaskets are designed to withstand temperature upto 110 °K.

5.8.6 Setting Blocks

Setting blocks are used to provide support in the relation to the size of glass, glazing techniques and condition of use. Setting blocks, which are rot-proof, non-absorbent and load bearing, capable of maintaining the requisite edge clearance without presenting local areas of stress to the glass through being incompressible or non resilient should be used.

5.8.7 Location Blocks

Location blocks are used between the edges of the glass, other then at the bottom edge, to prevent movement of the glass within the frame as the window or door is opened or closed and to prevent the weight of the glass causing the frame to become out of square. Location blocks should be of resilient non-absorbent material, generally of plasticized PVC and should be at least 25 mm long for all opening windows.

5.8.8 Distance Pieces

Distance pieces are made of a resilient, non-absorbent material, such as plasticized PVC and are placed to coincide on opposite sides of the sheet or unit at bead fixing points. These are ideally spaced at 300 mm centers and within 50 mm from the corners. Avoid placing them to coincide with setting or location blocks.

Innovative systems of glazing to fix the glass in buildings as well as important components of glazing have been described in this chapter with a view to assist the designer to adopt a suitable system of glazing as per his requirements.

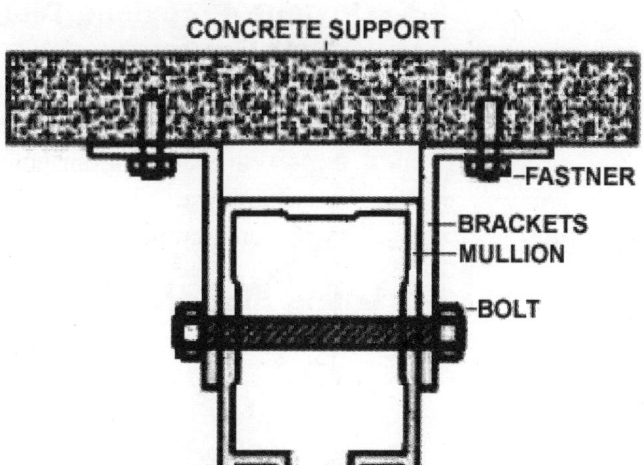

FIG. 5.1 CURTAIN WALLING - SUPPORT CONNECTION DETAIL

Dimensions (mm) change with specific conditions of building facades.

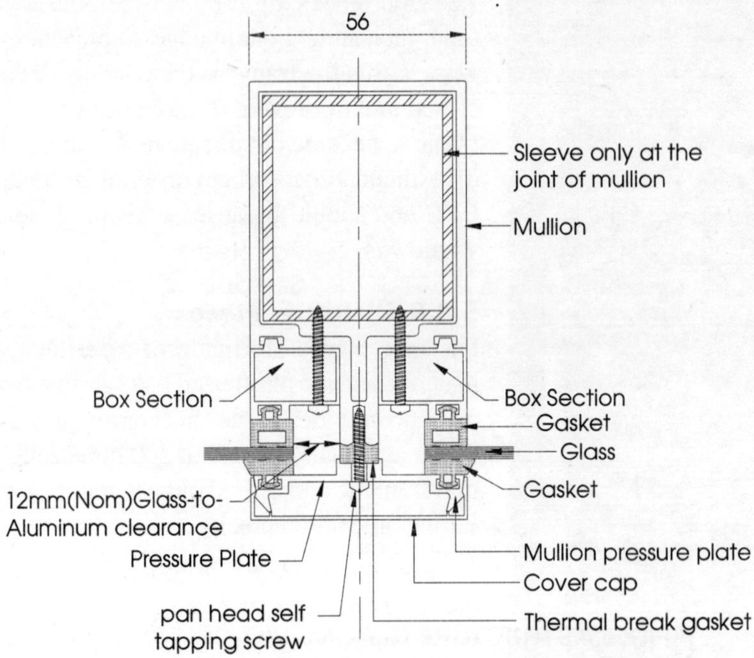

FIG. 5.2 CURTAIN WALLING - GLASS FIXING DETAIL

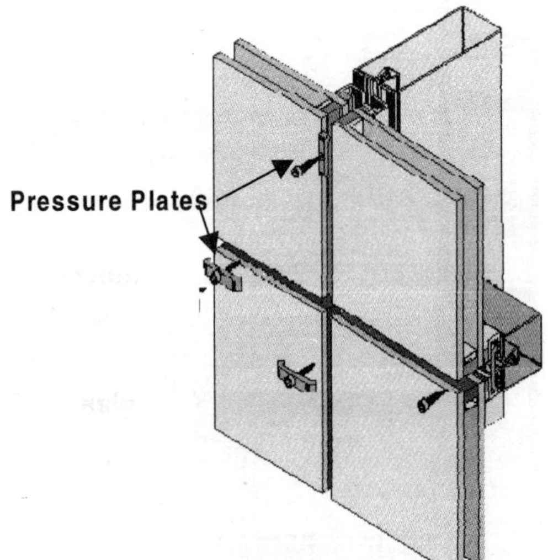

FIG. 5.3 CURTAIN WALLING - PRESSURE PLATES

Dimensions (mm) change with specific conditions of building facades.

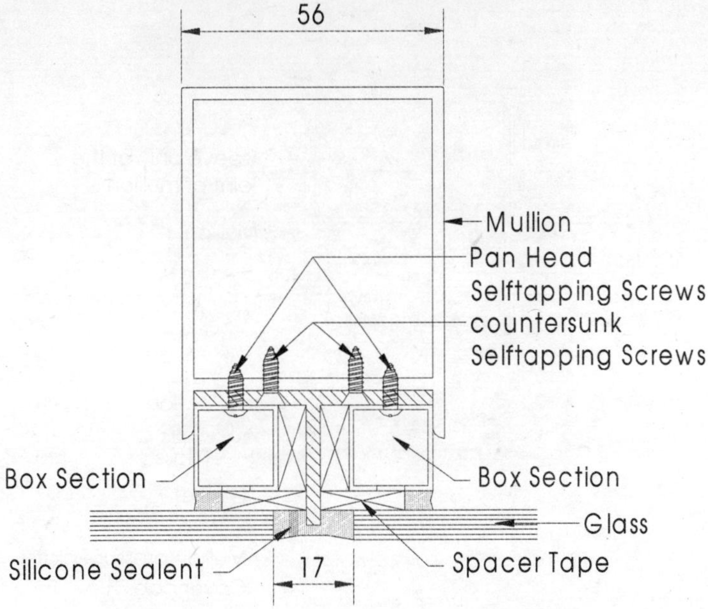

FIG. 5.4 STRUCTURE GLAZING DETAIL

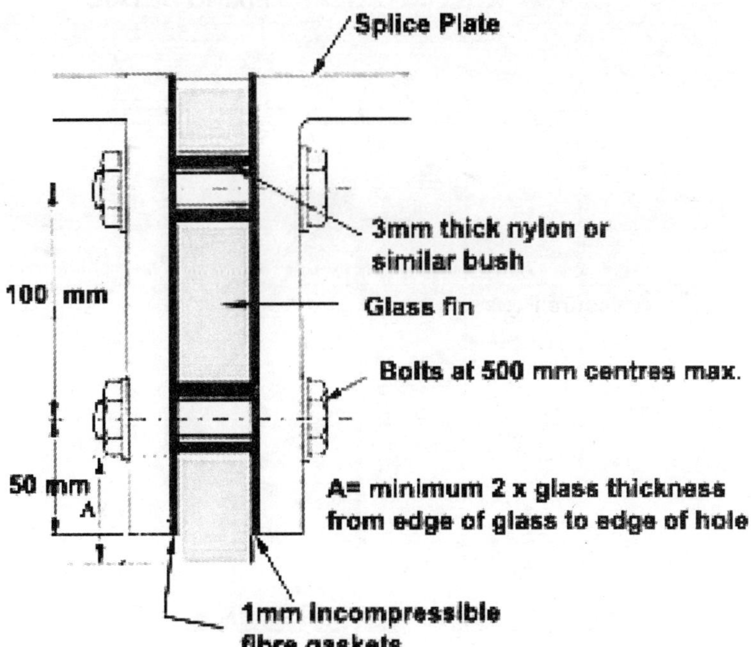

FIG. 5.5 BOLTED CONNECTION

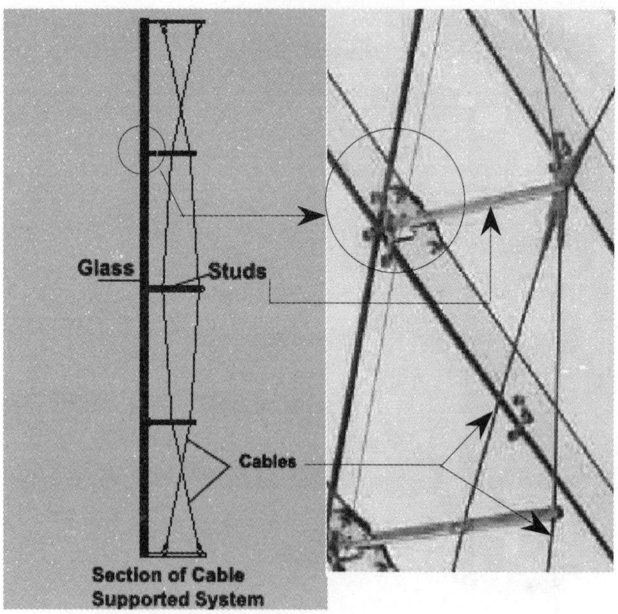

FIG. 5.6 CABLE STAYED GLAZING

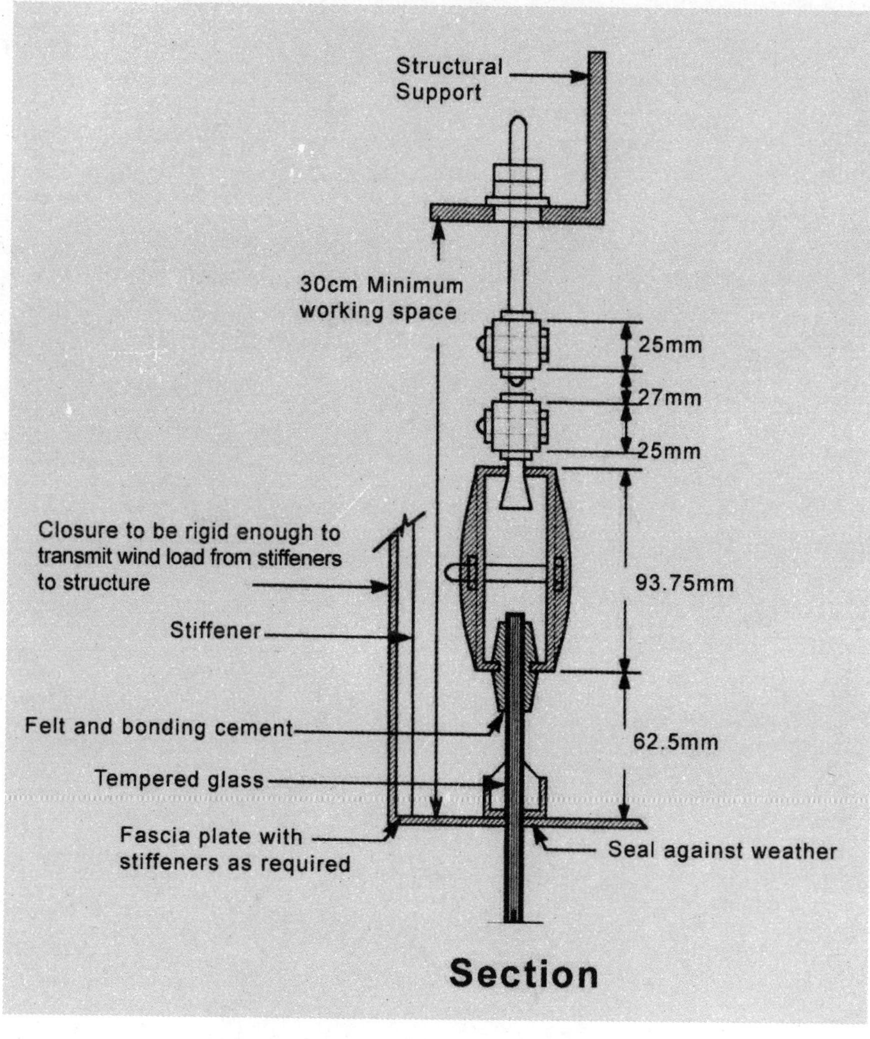

FIG. 5.7 SUSPENDED GLAZING

6

General Guidelines

6.0 GENERAL GUIDELINES

6.1 GENERAL

The following guidelines are for proper installation of glass in buildings:

- Windows or doors must not be operated till the glazing components are installed and the entire glazing system is complete.
- Glazing must never be done at temperatures below 4.44°C unless precautions are taken to prevent moisture where glass is to be installed.
- For safety and security from vandalism, tempered / laminated glass should be preferred in windows up to a height of 8-10 meters.

- There should not be any contact between glass and the frame.
- The performance of associated fittings & fixtures such as handles, bolts etc. should also be considered along with the strength of the glass with a view that if the human body strikes against the glass, while operating the fitting & fixtures, the safety of the human body is ensured. Under such situations tempered glass should be preferred.
- Details of installation vary with the type of material into which glass is to be set. This includes type of glazing material, gaskets, rope, tape, and glazing clips necessary for setting blocks, glazing compounds and sealants etc. Relevant usage details should be followed for each installation.

6.2 NORMAL (ANNEALED) GLASS

- Normal glass used for glazing should be free from all defects. However, if waviness is present, it should be adjusted horizontally to minimize optical distortion.
- When using normal / tinted glass, a check should be made that there is no variation in colours among various batches of glass as this spoils the visual continuity.

6.3 LAMINATED GLASS

- The edges of a laminated lite must be grounded and not be exposed to any prolonged contact with moisture. Consideration must be given to the use of weep holes or other alternate glazing procedures to ensure a dry framing cavity.
- Laminated glass is made by joining two or more sheets of glass with a plastic medium, mostly Poly Vinyl Butyral (PVB) between the sheets. PVB itself adheres well to glass. Laminating can also be done with certain resins. Some suppliers use PVC instead of PVB or use cheap resins which are difficult to make out at the time of supply but these materials will most certainly discolour and delaminate in a short period and such glasses will have to be changed. Precaution should be taken for right product.
- To protect human beings from injury, any facades more than 10° inclination should have laminated glass. In case of fall, the laminated glass will not break and if it breaks, it will break safely.
- When selecting laminated glass made from reflective glass, manufacturers should be consulted on methods of application.

6.4 TEMPERED GLASS

- The manufacturers should be checked for the availability of desired sizes as this sizes of tempered glass are custome mode.
- All door sizes, transoms, jambs, heads, locks, and any other special requirements must be detailed and all types of hardware must be selected before tempered glass is ordered, as its size can not be altered.

6.5 REFLECTIVE GLASS

- For general purpose cleaning use a mild fast-drying cleaning solution e.g. ammonia and water.

- The reflective coating should be on face II for better aesthetic, ease in cleaning & maintenance and to minimize external glare.
- Frequent glass cleaning during construction is recommended. Immediately remove any glazing lubricants from the reflective coated surface. Prolonged contact of the reflective coating with some glazing lubricants can cause damage to the coating.
- Household detergents should not be used as glass cleaning aids when reflective glass is used.
- Avoid any contact of the reflective coating with metals or other hard materials e.g. razor blades, belt buckles, buttons, rings or any abrasive clothing.
- Remove cement and other construction debris with a heavy rinse of water before using a squeegee or cloth.
- Clean a small area at a time and inspect the glass surface frequently to ensure that no glass or reflective coating is damaged. Glass should be covered with plastic film during the construction.
- Avoid contact of the reflective coating or glass with acids or strong alkalis. Substances such as caustic soda used to clean aluminum framing as it will cause extensive damage to the reflective coating and glass surface.
- Clean the glass when the surface is shaded. Abrasive cleaners, fluoride salts and hydrogen fluoride should not be used.
- Solvents, which may be used in moderation on reflective glass, include isopropyl alcohol, acetone, toluene and mineral spirits. Sample a small area prior to proceeding and immediately rinse the glass or coated surface with water to remove all traces of the solvent.
- When selecting reflective glass for a curtain wall system, all other types of glass components should match the glazing glass selected. It is advisable that full-size sections of the system be constructed for observation.

6.6 INSULATING GLASS

- Regardless of edge construction, the insulating edge seal cannot be exposed to moisture for prolonged time periods.
- Consideration must be given to weep holes or other alternate methods, which will assure a dry framing cavity.
- Handle and install insulating glass with care. Damaged edges and/or corners can result in breakage later. Insulating glass units should not be rolled on corners.
- The glazing compound must be a non-hardening type that does not contain any materials, which will attack the metal-to-glass seal of the insulating glass. Putty should never be used.
- Openings into which insulating glass is to be installed must be square and plumb. It is necessary to check that they are correct in size to meet the clearances necessary for the type of insulating glass being installed because insulating glass cannot be changed in size once it has been manufactured.
- Insulating glass must not have any areas covered with paint or paper because this can cause a heat trap that may result in breakage.
- There should be no direct contact between the insulating glass and the frame into which it is installed.

6.7 MIRROR

- Care should be taken not to scratch, chip or damage paint backing of the mirror because once the mirror paint is scratched, chipped or damaged, it becomes a weak point for moisture to enter and ultimately gives way to black spots which grow with time.
- While moving or storing the mirror, ensure that painted surface does not come in contact with any hard material as these hard materials may damage the mirror backing.
- If the mirror is to be installed in high humidity area like bathrooms, the edges should be sealed to prevent

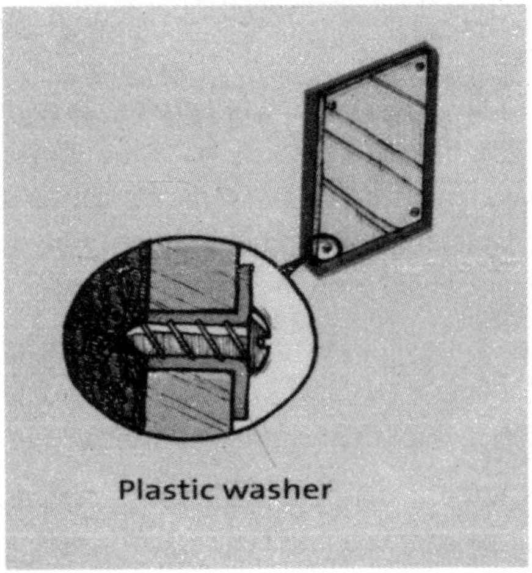

the moisture coming in contact with the mirror backing thus increasing the life of the mirror.

- Only neutral silicones or special sealants recommended for fixing mirrors should be used as adhesive where necessary. However, as far as possible, mirrors should not be pasted on the wall / plywood.
- Any adhesive containing acid like acetoxy silicones will damage the mirror. Adhesives like Fevicol or Araldite should never be used as their strong chemicals attack the mirror backing paint thereby resulting in black spots. Instead tapes can be used for fixing mirrors to the plywood. These double-sided tapes also form a barrier to moisture thus preventing the mirror from coming in contact with the moisture that may seep in from the sides.
- The surface of the wall / plywood on which mirror is installed should be dry as a wet surface either due to seepage, pipe leakage or humidity can damage the mirror.
- If mirror is to be fixed to a freshly built surface like newly plastered wall or over Plaster of Paris, care should be taken that the surface is properly cured, as any amount of moisture or dampness will damage the mirror.
- While cutting the mirrors, care should be taken not to chip the backing paint, as it becomes a weak spot for the moisture to seep in and damage the mirror backing.
- When laying the mirror down to cut or fabricate, ensure that the surface is clean. Felt is a recommended covering to prevent damage of mirror backing.
- Between the mirror and the mounting surface (wall, plywood etc.) a gap of 6mm to 10mm (for mirror of size more then 1 square meter) should be provided to ensure ventilation against humidity. The edges should be properly sealed.
- The mounting surface can be wall, plywood or asbestos sheet. Sheets of tin or plastic or felts or polyethylene sheets can be used between the plywood or asbestos and mirror.

- Proper thickness of mirrors should be used, especially if the mirror is to be installed in a vertical position. Mirror with less thickness may sag in the middle, resulting in distortion of image.
- Proper installation of mirrors requires that in almost all the cases the weight of the mirror must be supported at the bottom.
- Wall surfaces upon which mirrors are to be applied must be primed and sealed, smooth and firm, and thoroughly dry.
- The mirror should be mounted on a perfectly flat, clean and dry surface free from acids and aggressive substances. The mounting surface can be wall or plywood.

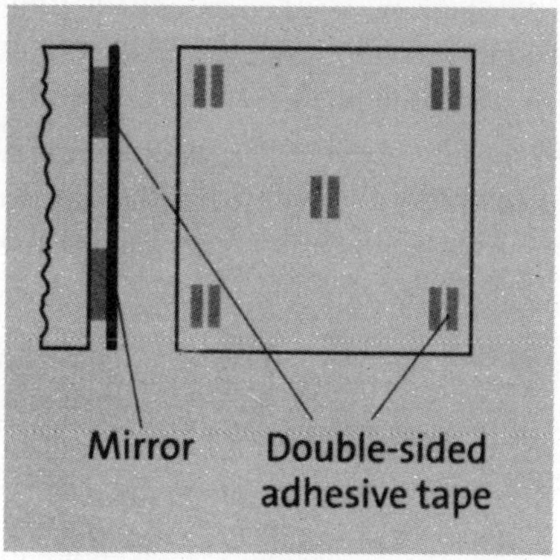

- For fixing of sheets at several levels one above the other, provide a space of 10 mm at the top and at the bottom for circulation of air.
- During the fixing of several mirrors side by side, leave minimum interleaving space (1 to 2 mm).
- In case of fixing of the mirror in a profile (frame), take care to see that the frame is perfectly clean and dry. Place the mirror supported on the hard non-metallic/ plastic wedges of at least 3 mm, to raise the mirror and thus avoid contact with the condensed water, which could accumulate in the profile.

6.8 ALUMINIUM FRAMINGS

- Thickness of powder coating: 60to80 micron
- Thickness of anodizing: 15 micron minimum
- Thickness of structural aluminum members including mullion, pressure plate, transom, structural frames and window frames in façade should be minimum 2mm so as to accommodate at least two threads of the screw. It is also recommended to have thickness of 2.5 mm at the point of screw fixing.

- The thickness of the cleats should be minimum 3mm.
- For sliding/ hinge windows minimum thickness of the frame should be 1.3mm.
- Thickness of any other non-structural members like cover plate, flashing etc should be minimum 1.1mm.

6.9 SAFETY FROM EARTHQUAKES

Preformance of window glass panels in the event of earthquakes is important for ensuring the safety of life as well as integrity of the building envelope. In plane deformation of the window frames is the primary cause of window glass damage. The deformation is accommodated by two consecutive mechanisms: rigid body motion of the glass panel in the window frame & the diagonal shortening of the glass under in-plane compressive forces. The first mechanism depends on the clearance between the glass panel & window frame and resiliency of the sealant material. The second mechanism is related to the mechanical properties of the glass panel. For greater safety from earthquakes:

- The edges of glass panels shall have adequate clearance from the frame (Top: 14 mm; Bottom: 10mm; left/right 12mm).
- The corners of glass panels shall be 'rounded' by removal of material at the corners and subsequent finishing of glass edges in the modified corner regions to minimize protrusions & edge surface roughness as well as to increase the dynamic displacement magnitude.
- Laminated glass should be used in view of its inherent properties.

General guidelines for installation as well as handling of glass have been discussed in this chapter for successful glazing operation in buildings

Reference Material

REFERENCE MATERIAL

1. Acoustics and noise control, 2nd edition by B J Smith, R J Peters and S Oween, 1996.
2. AS 1288 Australian Standard – 1994.
3. ASTM C 1279 Test method for Non- Destructive Photoelastic Measurement of Edge Surface Stresses in Annealed, Heat-Strengthened and Fully Tempered Flat Glass.
4. ASTM E 1300 US Standard
5. ASTMC 1036: C 162 Terminologies of Glass and Glass Products.
6. Australian Standard, Glass in buildings – Selection and installation
7. British Standard, BS 6262. Code of practice for glazing for buildings, Part 2. Heat, light and sound, 97/102384 DC.
8. British Standard, BS 6262. Code of practice for glazing for buildings, Part 6. Special applications. 97/102385 DC.
9. British Standard, BS 6262: Part 3 Code of practice for glazing for buildings. Part 3. Fire, security and wind loading, 96/100000 DC.
10. British Standard, BS 644. Wood windows. Part 1, Specification for factory-assembled windows of various types 98/103393 DC.
11. British Standard, Bullet-resistant glazing. Specification for glazing for interior use, BS 5051 –1:1988.
12. British Standard, Code of practice for installation of security glazing. BS 5367: 1995.
13. British Standard, Flat Safety Glass and Safety Plastics for Use of Buildings, BS 62O2 91
14. British Standard, Glass for glazing. Classification, BS 952-1: 1995.
15. British Standard, Glass for glazing. Terminology for work on glass, BS 952-2: 1980.
16. British Standard, Glass in Building – Laminated Glass and Laminated Safety Glass – Appearance, BS EN ISO 12543-6 98
17. British Standard, Glass in building Laminated glass and laminated safety glass. Definitions and description of component parts, BS EN ISO 12543-1:1998
18. British Standard, Glass in building, Security glazing, Testing and classification of resistance against explosion pressure. BS EN 13541: 2001
19. British Standard, Glass in building. Laminated glass and laminated safety glass, Dimensions and edge finishing, BS EN ISO12543
20. British Standard, Glass in building. Security glazing. Testing and classification of resistance against bullet attack, BS EN 1063: 2000.
21. British Standard, Glass in buildings. Insulating glass units. Part 1: Generalities and Dimensional tolerances.prEN1279-1) 95/105293 DC.
22. British Standard, Guide for security of buildings against crime, Warehouses and distribution units, BS 8220-3:1990.

23. British Standard, Guide for security of buildings against crime. Offices and shops, BS 8220-2: 1995.

24. British Standard, ISO/DIS 11600, Building construction Sealant, Classification and requirements, 00/102817 DC

25. British Standard, Performance of windows. Specification for operation and strength characteristics BS 6375-2: 1987

26. British Standard, prEN 12758-1, Glass in building. Glazing and airbome sound insulation. Part 1, Definitions and determination of properties 97/101808 DC.

27. British Standard, prEN 1279-4, Glass in building. Insulating glass units Part 4. Methods of test for the physical attributes of end seals.96/106242 DC

28. British Standard, prEN 13022-1. Glass in building. Structural sealant glazing. Part 1. Actions, requirements and terminology, 97/107906 DC.

29. British Standard, prEN 13024-1. Glass in building. Thermally toughened borosilicate safety glass. Part 1, Specifications, 97/107905 DC

30. British Standard, prEN ISO 12543-2 Glass in building. Laminated Glass and laminated safety glass Part 2: Laminated safety glass (ISO/DIS 12543-2) 95/102177 DC

31. British Standard, prEN ISO 12543-2. Glass in buildings. Insulating glass units. Part 3, Initial type testing on gas-filled insulating glass units, Gas leakage rate. 95/104639 DC

32. British Standard, prEN ISO 14439 Glass in building. Glazing requirements. Glazing blocks (ISO/DIS 14439:199595/101448 DC

33. British Standard, Specification for aluminum alloy framed sliding glass doors 90/13950 DC

34. British Standard, Specification for aluminum framed sliding glass doors, BS 5286:1978.

35. British Standard, Specification for anti-bandit glazing (glazing resistant). BS 5544: 1978

36. British Standard, Specification for impact performance requirements for flat safety glass and safety plastics for use in buildings.95/107225 DC

37. British Standard, Specification for linseed oil putty for use in wooden frames, BS 544: 1969

38. British Standard, Windows, doors and roof-lights, Code of practice for safety in use and during cleaning of windows and doors (including guidance on cleaning materials and methods), BS 8213-1: 1991.

39. British Standard, Workmanship on building sites. Code of practice for glazing, BS 8000-7: 1990.

40. BS 5516 : 1991 Code of Practice for design and installation of sloping and vertical patent.

41. BS 6206: 1981 Specification for impact performance requirements for flat safety glass and safety plastics for use in buildings.

42. BS 6262-4: 1994 Glazing for buildings. Safety related to human impact.

43. BS EN 1096-1: 1999 Glass in building. Coated glass. Definitions and classification.

44. BS EN 1634-1: 2000 Fire resistance tests for door and shutter assemblies Fire doors and shutters.

45. BS EN 1748-2: 1998 Glass in building. Glass ceramics.
46. BS EN 572-2: 1995 Glass in building. Basic soda lime silicate glass products. Float glass.
47. BS EN ISO 12543 Glass in building. Laminated glass and laminated safety glass. Test method for durability.
48. Construction Materials, Types, users and application, Second Edition, Caleb – Hornbostel, John Wiley & Sons, 1991
49. DIN 1249 81 teil 1 81 Glass in Buildings: Sheet – Glass, Concepts Dimensions.
50. DIN 1249 87 teil 10 81 Glass in Buildings: Chemical and physical Properties.
51. DIN 1249 blatt 1 Flat Glass for Building Construction. Window Glass.
52. DIN 18516 Part 4 Mechanical properties of Glass: Minimum flexural strength 75 N/ sq. mm under tension 120 N/sq. mm under compression.
53. Glass in building- insulating glass units. Part 1. Actions, requirements and terminology, 97/ 107906 DC prEN 1279-1 96.
54. Glass technology live (Architectural Exhibition) Proceeding of Architektur Sonderausstellung Glasstec 2002 messe Dqsseldory 28.10 – 1.11.2002 Hallen 10+11.
55. Guardian Industries 1992 (Unpublished Compilation)
56. Handbook of Glass in Construction by Joseph S. Anstock, McGraw- Hill Publication, 1997.
57. Indian Standard, Code of Practice for design Loads (other than Earthquake) for Buildings and Structures, Part 3 wind loads (Second Revision) Sixth reprint Nov. 1998.
58. Indian Standard, Code of Practice for fixing and Glazing of Metal (Steel & Aluminium Doors, windows & Ventilators), IS- 1081- 1960.
59. Indian Standard, Code of Practice for Glazing in Buildings, IS- 3548-1988.
60. Indian Standard, Code of Practice for Patent Glazing, IS- 10439 – 1983.
61. Indian Standard, Code of Practice for Sound insulation of Non Industrial Buildings, IS- 1950-1962.
62. Indian Standard, Code of Practice for Structural Safety of Buildings Loading Standards, IS- 875-1964.
63. Indian Standard, Code of Practice for Structural Safety of Buildings Loading Standards, IS- 1642-1960.
64. Indian Standard, Glossary of Terms relating to glass and glassware, IS- 1382 – 1981.
65. Indian Standard, Guide for day lighting of buildings, IS- 2440-1975.
66. Indian Standard, Solar Flat Plate Collector Specification, IS- 12933-1992
67. Indian Standard, Specification for flat Transparent sheet glass, IS- 2835 – 1977.
68. Indian Standard, Specification for Steel Doors, Windows and Ventilators, IS- 1038.
69. Indian Standard, Specification for Steel Glass Front Cabinets, IS – 7760.
70. Indian Standard, Specification for Transparent sheet glass for glazing and framing purposes, IS- 1761-1960.

71. Indian Standard, Specification for Wired and Figured Glass, IS- 2553-1971.

72. Indian Standard, Specification for wired and figured glass, IS- 5437- 1969.

73. Indian Standard, Timber Paneled and Glazed Shutters Specification, IS- 1003 Part 2.

74. Indian Standard, Timber Paneled and Glazed Shutters Specification Door shutters, IS- 1003 Part 1.

75. Indian Standard, Transparent Float Glass – Specification, IS- 14900-2000.

76. Indian Standard, Wind Code, IS: 875 (Part-3)-1987(reaffirmed 1997).

77. Laminated Architectural Glass, Specification guide 1998 – Saflex.

78. Modiguard, Reflections Volume II, Issue 2, 2000.

79. Modiguard, Reflections Volume III, Issue 2, 2001.

80. Modiguard, Reflections Volume III, Issue 3, 2002.

81. Modiguard, Reflections Volume III, Issue I, 2001.

82. prEN 12600. Glass in building. Pendulum test Impact test method for flat glass and performance requirements 99/108002 DC

83. prEN 1279-1 96, Glass in building Insulating Glass Units, Part 1 Generalities Dimensional tolerances and rules for the system description,

84. prEN 1279-296 Glass in Building – Insulating Glass Units, Part 2 Long term method and requirements for moisture penetration.

85. prEN 13022-2 Glass in building, Structural sealant glazing. Part 2, Glass, 97/107907 DC.

86. prEN 13022-2 Glass in building, Structural sealant glazing. Part 3. Sealants. Test methods. 97/107908 DC.

87. prEN 1347-1. Glass in building. Design of glass panes Part 1, General basis of design 99/100074 DC

88. PrEN ISO 12600 Glass in Buildings, providing performance criteria for the type of glass use in building

89. Rubber, vulcanized and thermoplastic. Performed gaskets used in buildings. Classification, specifications for materials and test methods for gaskets (ISO/DIS 3934). 98/126150 DC.

90. Specification for aluminum alloy windows 98/103393 DC.

91. Technical Manual, user's guide SOLARBEL.

92. TH 42119 Glass in building. Windows, Doors and Furniture.

93. TH 42119 Glass in building. Windows, Doors and Furniture.

94. Trutuf, High Performance Glass Products, GSC Toughened Glass Pvt. Ltd. (Company Brochure).

Definitions

DEFINITIONS

Ambient Noise - The all-encompassing noise associated with a given environment, usually a composite of sounds from sources near and far.

Ambient Temperature - The environment temperature surrounding the object.

Annealing - The process which prevents glass from shattering after it has been formed. The outer surfaces of the glass shrink faster than the glass between the surfaces, causing strain which can lead to shattering. Reheating the glass and allowing it to cool slowly can avoid this.

Aspect Ratio - The ratio of the longer side of a panel to its shorter side.

Balustrade - A low wall forming a parapet to a stair, ramp, balcony, raised level, or a change in levels.

Blocks - Rectangular, cured sections of neoprene or other approved materials, used to position a glass product in a glazing channel.

Bronze Glass - A glare – or heatreducing glass intended for applications where glare control and reduction of solar heat are desired or where color can contribute to design.

Chair Rail - A fixed glazing bar or rigid push bar, which provides protection from human impact.

Clear Glass - As its name states, transparent or clear glass.

Clips -Wire spring devices to hold glass in a rabbet sash, without stops or face glazing.

Coating - A material, usually liquid, used to form a covering film over a surface. Its function is to decorate and/or protect the surface from destructive agents or environments (abrasion, chemical action, solvents, corrosion, and weathering).

Decibel (dB) - A unit adopted for convenience in representing vastly different sound pressures. It is 20 times the logarithm to the base 10 of the ratio of the sound to the pressure of 0,0002 dyne/cm^2. This reference pressure is considered the lowest value that the ear can detect.

Deflection - The degree of inward or outward movement in lites of glass exposed to unequal pressures from either side of the glass. This condition may be permanent or temporary. Deflection can be caused by the wind or temperature, elevation, or barometric pressure changes. Air-absorbing desiccants can contribute to deflection.

Delaminate - To split a laminated material parallel to the plane of its layers. Sometimes used to describe cohesive failure of an adherent in bond strength testing.

Desiccants - Porous crystalline substances used to absorb moisture and solvent vapors from the air space of insulating glass units. More properly called absorbents.

Door - A hinged, sliding, or otherwise supported openable barrier providing entrance to and exit from a building, corridor, or room. Doors may be framed or unframed.

Double Glazing - Glazing that incorporates two panels, separated with an air space for the purpose of sound insulation or thermal insulation, or both.

Emissivity - The relative ability of a surface to radiate heat.

Fin - A piece of glass positioned to provide lateral support.

Flame Retardant - Substances mixed with rubber to retard its burning (e.g. highly chlorinated hydrocarbons). Neoprene is less flammable than natural rubber.

Flat Glass - Pertains to all glasses produced in a flat form.

Float Glass - Transparent glass with flat, parallel surfaces formed on the surface of a pool of molten tin.

Frequency - The number of times an action occurs in a given time period. In sound, the number of complete vibration cycles per sound, represented by the hertz (Hz).

Frame - A structure manufactured from timber, metal, glass, or other durable material or combinations of materials, such as glass fins and structural sealant supporting the full length of all the edges of the glazed panel.

Gasket - Pre formed shape, such as a strip, government etc., of rubber and rubber-like composition used to fill and seal a joint or opening, alone or in conjunction with the supplemental application of a sealant.

Glass - A hard brittle amorphous substance produced by fusion and usually consisting of mutually dissolved silica or silicates that also contain soda and lime. It may be transparent, translucent, or opaque.

Glazing - The securing of glass in prepared opening in windows, door panels, screens, partitions, etc.

Heat-Absorbing Glass - Glass (usually tinted) formulated to absorb an appreciable portion of solar energy.

Heat-Strengthened Glass – Glass, which has been subjected to special heat treatment so that the residual surface Compression stress and the edge compression stress lies between those of ordinary annealed glass and toughened glass.

Intensity - The rate of sound energy passing through a unit area. The intensity is measured by the energy in ergs transmitted per second through 1 cm^3 of surface. The energy in ergs per cm^3 in a sound wave is given by $E = 2\grave{A}^2 dn^2a^2$.

Interlayer - The transparent damping material used in laminated glass.

K-Value - The European equivalent of the American (ASHRAE) U-Value. The Units are W/m2K and are based on a wind speed of 4.4 m/sec at O°C with an indoor temperature of 20°C.

Light Transmittance - Clear glass, depending on its thickness, allows 75 to 92 percent of visible light to pass through.

Neoprene - A synthetic rubber with physical properties closely resembling those of natural rubber but not requiring sulfur for vulcanization. It is made by polymerizing chloroprene, and the latter is produced from acetylene and hydrogen chloride.

Non-Residential Buildings - Buildings such as hotels, hostels, motels, shops, offices, schools, public assembly buildings, and factories, and those parts of residential buildings common to a group of dwellings such as common circulation areas in blocks of two or more flats.

Patterned Glass - Rolled glass having a distinct pattern on one or both surfaces.

Rabbet - A two side L-shaped recess in sash or frame to receive lites or panels. When no stop or molding is added, such rabbets are face-glazed. Addition of a removable stop produces a three-sided, U-shaped channel.

Reflective Coated Glass - Glass with metallic or metallic oxide coatings applied onto or into the glass surface to provide reduction of solar radiant energy, conductive heat energy, and visible light transmission.

Rw – Weighted sound reduction index.

Sash -A frame into which glass products are glazed, i.e., the operating sash of a window.

Sealant - A material used to fill a joint, usually for the purpose of weatherproofing or waterproofing. It forms a seal to prevent gas and liquid entry.

Setting Blocks - Small blocks of composition, lead, neoprene, wood, etc., placed under the bottom edge of the lite or panel to prevent its settling onto the bottom rabbet or channel after setting, thus distorting the sealant.

Shading Coefficient - The ratio of the rate of solar heat gain through a specific unit assembly of glass to the solar heat gain through a single lite of 3 mm clear glass in the same situation.

Skylight - A glass and frame assembly installed into the roof of a building.

Sloped Glazing - Any installation of glass that is at a slope of 15° or more from the vertical.

Solar Energy Absorption - The percentage of the solar spectrum energy (ultraviolet, visible, and near-infrared) from 300 to 300 nm that is absorbed by a glass product.

Solar Energy Transmittance (Direct) - The percentage of energy in the solar spectrum, ultra-violet, visible, and near infrared energy, 300 to 4,000 nanometers, that is directly transmitted through the glass.

Sound Absorption - The property possessed by material and objects, including air, of converting sound energy to heat energy.

Spacers - Spacers are used to prevent displacement of the glazing sealants by wind pressure on the glass.

Span - The dimension between supports. For panels supported on all four edges, it corresponds to the smaller of the sight size dimensions.

Spandrel - That portion of the exterior wall of a multistory commercial building that covers the area below the sill of the vision glass installation.

Structural Glazing - Structural glazing is a method of bonding glass and insulating glass units to an aluminum window frame or curtain wall system utilizing a high-strength, high-performance silicone sealant.

Suspended Glazing - Suspended glazing is a method of providing a frame less glazing façade to fix together a matrix of toughened glass lites, hung from the building structure.

Tempered Glass - Fully tempered (FT) glass is reheated to just below the softening point (about 1300°F or 704.4°C) and then rapidly cooled.

Thickness - The dimension between the two surfaces of a panel of glass, other than its length or width.

Tinted Glass - Body-colored glass with specific ingredients formulated to produce light reducing and/or heat absorbing glass products.

Toughened Glass - Glass that has been subjected to special heat or chemical treatment so that the residual surface compression stress and the edge compression stress is greater than heat-strengthened glass.

Transmission - The passage of sound from one location to another through an intervening medium, such as a partition or air space.

Transmittance - The fraction of radiant energy that passes through a given material.

Ultraviolet Light (UV) - A form of luminous energy occupying a position in the spectrum of sunlight beyond the violet and having wavelengths of less than 3900 A, which is the limit of the visible spectrum. Ultraviolet rays are very active chemically, exhibit bactericidal action, and cause many substances to fluoresce.

UV Transmittance - The percentage of energy in the ultraviolet (UV) spectrum, from 300 to 380 nanometers, that is directly transmitted through the glass.

Visible Light Transmittance - The percentage of light in the visible spectrum, from 380 to 780 nanometers, that is transmitted through the glass.

Wind Load - Load on glass because of the speed and direction of the wind.